ARCHITECTURAL ROBOTICS

ARCHITECTURAL ROBOTICS

ECOSYSTEMS OF BITS, BYTES, AND BIOLOGY

KEITH EVAN GREEN

THE MIT PRESS
CAMBRIDGE, MASSACHUSETTS
LONDON, ENGLAND

© 2016 Massachusetts Institute of Technology

All rights reserved. No part of this book may be reproduced in any form by any electronic or mechanical means (including photocopying, recording, or information storage and retrieval) without permission in writing from the publisher.

This book was set in PF Din Pro by The MIT Press. Printed and bound in the United States of America.

Library of Congress Cataloging-in-Publication Data

Names: Green, Keith Evan, author.
Title: Architectural robotics : ecosystems of bits, bytes, and biology /
Keith Evan Green.
Description: Cambridge, MA : The MIT Press, 2016. | Includes bibliographical references and index.
Identifiers: LCCN 2015037895 | ISBN 9780262033954 (hardcover : alk. paper)
Subjects: LCSH: Architecture and technology. | Architecture—Human factors. |
Cooperating objects (Computer systems) | Robotics—Human factors. |
Buildings—Environmental engineering. |
Smart materials in architecture. |
Intelligent buildings.
Classification: LCC NA2543.T43 G69 2016 | DDC 720/.47—dc23
LC record available at http://lccn.loc.gov/2015037895

10 9 8 7 6 5 4 3 2 1

No one can predict what environments architecture will create. It satisfies old needs and begets new ones.

—Henri Focillon, *The Life of Forms in Art*, 1934

We believe it's technology married with the humanities that yields us the results that make the heart sing.

—Steve Jobs, launch of iPad 2, San Francisco, March 2, 2011

CONTENTS

Preface ix
Acknowledgments xi

PROLOGUE
1. SCENES FROM A MARRIAGE 1
2. OF METAPHORS, STORIES, AND FORMS 7

■ **RECONFIGURABLE**
3. SPACES OF MANY FUNCTIONS 23
4. IN AWE OF THE NEW WORKFLOW 41
5. THE RECONFIGURABLE ENVIRONMENT 68

■■ **DISTRIBUTED**
6. LIVING ROOMS 71
7. THE ART OF *HOME+* 85
8. THE DISTRIBUTED ENVIRONMENT 126

■■■ **TRANSFIGURABLE**
9. PORTALS TO ELSEWHERE 129
10. A LIT ROOM FOR THE PERSISTENCE OF PRINT 137
11. THE TRANSFIGURABLE ENVIRONMENT 170

EPILOGUE
12. ECOSYSTEMS OF BITS, BYTES, AND BIOLOGY 173

Notes 205
Bibliography 233
Index 249

PREFACE

This book is not about architecture or computing but architecture *and* computing, which might prove frustrating or at least disappointing to some readers who were seeking more attention to and validation of their own disciplines and practices. My 9-year-old twin children, Maya and Alex, girl and boy, know this disappointment and frustration far better than I do, and as the parent to both them and this book, I am acutely aware, the moment I attend to one, that I have left the other out of the conversation. Or, at least, I partly fail to keep them both at hand, doing my best to speak to both at once.

"Architecture encompasses everything," wrote Italian architect Giò Ponti. This is my challenge in writing this book. While *architectural robotics* is a narrow field of consideration, and while the body of work elaborated and otherwise referenced in these pages is in its infancy, the reach of this book is expansive: from architecture and robotics and human-computer interaction to the fine arts, philosophy, interaction design, industrial design and engineering, anthropology, education, medicine, psychology, cognitive science, and neuroscience. My formal education, life experiences, and street smarts get me through, but across these many pages I surely enter territory where I'm more stranger than native. Please accompany me here, recognizing my intention is to grapple with a different way of thinking about, designing, and engaging some hybrid of architecture and computing.

Clearly, the subfield of architectural robotics is rapidly emerging, as observed in the addition of new academic programs and researchers concerned with it and in the entry of industry and venture capital into the domain of smart products, intelligent building systems, and assistive robotics. It's difficult to tell whether this book comes too early or too late, but I hope that, more than anything else, it has an influence on the way designers, defined in the broadest sense, view our world of physical bits, bytes, and biology and imagine making a contribution to it.

ACKNOWLEDGMENTS

Many thanks are due:

- to Henriette Bier, Kas Oosterhuis, and their Hyperbody Research Group at the Technical University of Delft (TU Delft) and to Tomasz Jaskiewicz, Pieter Jan Stappers, and their ID-StudioLab, also at TU Delft, for providing me with a commodious place in the Netherlands to write much of this book.
- to Imre Horváth at TU Delft for energizing conversation and debate at the beginning and end of this process.
- to Ufuk Ersoy, Robert Benedict, Yixiao Wang, and many others at Clemson University, and to its College of Architecture, Arts & Humanities for supporting this book.
- to Mark D. Gross, director of the ATLAS Institute of the University of Colorado, Boulder, who collaborated with me on preparatory phases for this book, and who became a quiet, propelling force.
- to my family, who put up with me during another long haul.
- to the U.S. National Science Foundation for supporting each of the three case studies of this book, and especially to Ephraim Glinert, NSF program director, for his guidance and belief in an "architectural robotics future."
- to my Clemson graduate students and colleagues who made one of the three case studies of this book the primary focus of their research for several years, which qualifies each of them as a "coauthor":
  Henrique Houayek, Martha Kwoka, Lee Gugerty, and Jim Witte (AWE)
  Tony Threatt, Jessica Merino, Paul Yanik, and Johnell Brooks (ART)
  George Schafer, Amith Vijaykumar, and Susan King Fullerton (LIT ROOM/LIT KIT)
- to, especially, Ian D. Walker, professor of electrical and computer engineering, who was my research partner for all three case studies, who graciously edited the technical parts of this manuscript, and who continues to realize with me the ecosystems of architectural robotics.

PROLOGUE

1 SCENES FROM A MARRIAGE

Consider, from a skeptical distance, the scene evoked by the term *human-computer interaction* as if it were a play being performed in a theater. We have an empty stage but for a desk and a desk chair, both of no particular qualities. The lighting is nondescript. On the desk is a desktop computer, probably an IBM. (Is it a "386"?) And seated before the computer, "the human." The monitor of the computer is positioned at eye level of its user. The physical size of the computer appears to be not much larger than the size of the human's head, so that the human and the computer appear as intimates.

This is scene 1 of human-computer interaction, set in 1983, as portrayed in *The Psychology of Human-Computer Interaction*, the book that popularized the term *human-computer interaction* (or HCI, for short).[1] In the first figure appearing in this foundational book written by Card, Moran, and Newell (figure 1.1), a bubble-headed human, happily and without distraction, commands a CPU connected to the television-like computer display that was typical of its day. So engrossed is the merry human in his interactions with the desktop computer that he remains oblivious to the fact that his nose is practically one with the display (and may even be polishing it!). Observe, as well, that our bubble-headed friend is seemingly no more than a head and upper torso sharing a vacuous space with the computer. The desk chair in this scene is only signified. The desk and stage are absent altogether.

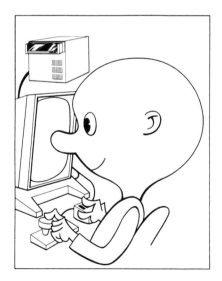

1.1
Scene 1: One human and one computer (from T*he Psychology of Human-Computer Interaction*, reprinted with permission of Taylor & Francis).

This scene 1 of human-computer interaction can hardly portray our engagement with computing in the new millennium. In scene 2, which begins in 2007 with the introduction of the iPhone, computing finds its way into our pockets, our cars, our appliances, ... into our everyday lives, indoors and out. This is *ubiquitous computing*, the kind of computing that is all around us, woven "into the fabric of everyday life," now that computer hardware is cheaper and more powerful, more miniaturized, more readily available, and more accessible to so many more of us.[2]

Mark Weiser's vision of ubiquitous computing is compelling for recasting the "one human–one computer" of scene 1 of human-computer interaction as a web of interconnected devices and people.[3] Still, to the skeptical audience, scene 2 can appear just as strange as scene 1—at least as presented in the foundational paper "Ten Dimensions of Ubiquitous Computing"[4] written by Steve Shafer of Microsoft Research. In the only graphic image included in this paper (figure 1.2), the "stage floor" is populated with a scattering of interconnected computer-hardware devices. There is not a sole human actor here but instead three actors—identical triplets, in fact—bubble-headed again, and delineated with less detail than that given to most of the devices. Note, also, that there is neither a desk nor a stage of any kind here. Not even a chair. Where are we? Do we envision ourselves empowered by this network of devices? Entangled by its web? "Alone together?"[5]

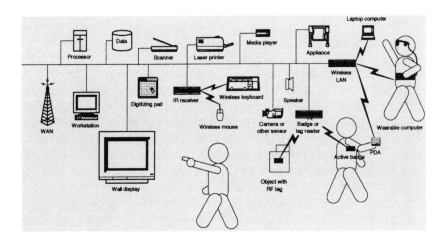

1.2

Scene 2: Computing all around us
(from "Ten Dimensions of Ubiquitous Computing," reprinted with permission of Springer-Verlag).

While such sobering questions surfacing from scene 2 merit a reply, the human-computer interaction of the very near future promises to be an altogether different performance: a scene 3 in which the distance between computers and humans approaches naught, figuratively and literally. Some of us might prophesy this scene as the most unsettling one yet. Amplifying the disquiet may be the recognition that a mere thirty years have passed between the settings of scenes 1 and 3.

In a presentation mode appropriate for the information age, scene 3 of human-computer interaction is composed of three concurrent videos, each presented on wall-sized computer displays:

On display 1: *Computing on us and inside us*
On display 2: *More of us in computing*
On display 3: *Computing embedded in our everyday living environments*

At the smallest physical scale, the computing shown on display 1 resides not only around us but also *on* us and *in* us. One instance of *computing on us* is the embedded, bionic second-skin forging a connection between our bodies and the external world. The "e-skin" prototyped by John Rogers can be stuck on or laminated directly to human skin, while that developed by Takao Someya, one tenth the thickness of plastic kitchen wrap, conforms to any moving body part.[6] Collectively, these bionic second-skins are soon expected to monitor our medical conditions and provide more sensitive, more life-like prosthetics.

Rogers and Someya are prominent members of a considerable community of researchers striving to situate *computing on us and inside us*. Meanwhile, the kind of computing shown on display 2 is also gaining momentum: *more of us in computing*. On display 2, what's in us—some aspects of our physical and mental selves that we might define as human—is coming to define the behavior of intelligent machines, extending the reach of computer software and hardware and, in turn, promising to help us learn more about ourselves. Humanoid robotics is the most visible case, the topic of countless lay-press features, particularly since its 2007 cover treatment by the *New York Times Magazine*.[7] When and if the *singularity* of machines and humankind will occur are questions debated passionately.[8]

This book is not focused on displays 1 and 2, but rather on display 3, where computing—specifically robotics—is embedded in the very physical fabric of our everyday living environments at relatively large physical scales ranging from furniture to the metropolis. While display 1 demonstrates computing at the smallest physical scale, and display 2 demonstrates computing at a physical scale that we can manipulate with our hands, display 3 demonstrates computing at a scale *larger than us*—a scale that, for us, becomes spatial in a physical

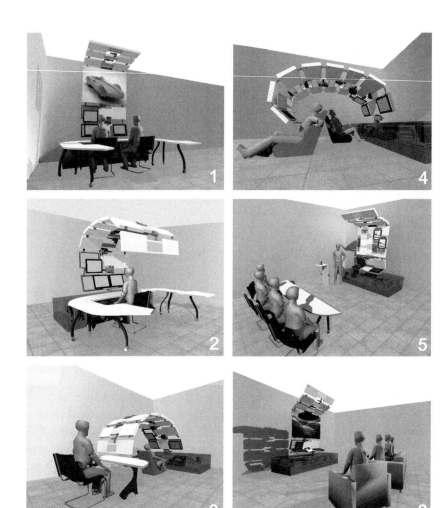

1.3
Scene 3, display 3: Computing embedded in our everyday living environments (the Animated Work Environment).
In each panel of this figure, the number shown in the lower-right corner (e.g., "1") denotes a particular physical confguration assumed by the Animated Work Environment, as elaborated in chapter 4.

way, matching, exceeding, or enveloping our physical bodies. Also discernible on display 3 is the unfolding of countless, everyday interactions across multiple human actors and things that are physical, digital, and/or biological: an *ecosystem of bits, bytes, and biology*.

One such sequence on display 3 presents working life in the near future, over the course of some number of hours (figure 1.3). At first, two people are shown collaborating on a written report, situating themselves about an extended table facing an upright, robotic wall of computer displays; sometime later, the same two people are seated back-to-back, working on different aspects of a joint project, enveloped by the robotic wall, only now configured as a womb-like enclosure that supports independent work and intermittent consultation. At any instance, the precise configuration assumed by the robotic work environment may be determined by any one of these possible paths:

- An actor selecting one of the six preprogrammed configurations.
- An actor tuning and saving a configuration for herself.
- The work environment assessing the activity of the actors and then configuring itself to accommodate the actors' activity.

If these events were to be communicated in print (in keeping with the black-and-white "set designs" of scenes 1 and 2 as presented in the previous figures), they would appear as a sequence of storyboard frames (figure 1.3) capturing discrete instances of human-computer interaction over time, populated by actors, alone or in collaboration, interacting with an environment that itself assumes different physical configurations as working activities demand.

This robotics-embedded work environment—more precisely, the Animated Work Environment that was prototyped in my lab—is an instance of architectural robotics, a *cyber-physical environment*, an environment characterized by the synergetic blending of computing and the physical realm (its materials, components, and systems).[9] Artfully and meticulously designed, architectural robotics is interactive, adaptive, and at least partly intelligent.

On display 3, we might also imagine a scene in which robotics embedded in the physical environment occurs at scales larger than this collaborative workstation. Here, architectural robotics extends from the theater stage itself, to the theater building that envelops the stage, to outside the confines of the theater walls, and into the infrastructure of the town or city where the theater is located.

Inevitably, the built environment will become increasingly interactive and intelligent. And given its scale and complexity, the built environment will provide the grand stage for the other HCI scenes so far considered: it envelops the desk and the desktop computer, and all the devices we carry and that exist around us, and all the human users empowered by and subjected to the vast web of computation. Consequently, any consideration of the emerging, cybernetic built environment must address, in a fundamental way, a different kind of human-computer interaction: as we come to live in and around computing, it arguably becomes us, and it inevitably comes to live within us and becomes our home.

2 OF METAPHORS, STORIES, AND FORMS

"Metaphors," writes historian Harry Francis Mallgrave, "are not just flourishes of language: they are the essential rudiments out of which we conceptualize ... the world."[1] Nobel laureate George Edelman calls them "a powerful starting point for thoughts that must be refined by other means."[2] In fewer words, philosopher José Ortega y Gasset considers metaphors a "tool for creation."[3] Not casually, the first chapter of this book relies on use of metaphor, where the terms of theater (*scenes, actors, stage*) and the terms of film and other time-based media (*storyboards, displays*) convey three milestones in human-computer interaction (HCI). For the purposes of the previous chapter, the theater metaphor captures, importantly, the transformative human-computer interaction of display 3, in which HCI becomes the interaction of people, things, and the physical environments that surround them—an *ecosystem of bits, bytes, and biology* (recognizing that *bits* are not only digital but also physical).

From this vantage, the theater is not simply home to artifice, any more than computing is an electronic abacus or architecture is its plan. Rather, the theater (as metaphor, and not) makes tangible the kind of empirical frame that probes the relationships across people and things, exhibited often enough by film, HCI, and architecture. Theater, film, HCI, and architecture share the capacity to cultivate an understanding of human experience; in this sense, they are, to borrow philosopher Don Ihde's term, "epistemological engines" for gaining knowledge about how we live our lives.[4] Moreover, theater, film, HCI, and architecture are frequently conflated in ways that are complex and constructive, and briefly examining their intersections offers a powerful starting point for recognizing the nature and promise of architectural robotics for an increasingly digital society.

Beginning at the intersection of theater and architecture, Milanese architect Aldo Rossi, for one, imagined the theater much as any other type of architectural work: on one hand, as a "tool, an instrument, a useful space where a definitive action can occur";[5] and on the other hand, where "people, events, [and] things ... continuously intersect."[6] However, for Rossi, what distinguishes the theater from the other typologies of architecture (e.g., schools, libraries, hospitals) is its accommodation of human action that is, at once, precisely scripted (by the playwright and director) and also open to interpretation and happenstance (of a given performance). For Rossi, the theater's way of weaving "strangely together" art and life—its "magical ability to transform every situation" familiar to us into something unexpected or at least compelling[7]—intrigued the Milanese architect and motivated his design activity with respect to not only his design of theater buildings but also his overall design *opera*. Recognizing the generative capacity of theater, Rossi characterized the theater building coupled with

2.1
The *Théâtre de l'espace*, based on Edouard Autant's plan and section (used with permission of Gray Read, Florida International University).

the performances it accommodates as, alternatively, a "theatrical machine," a "scientific theater," and a "laboratory" for observing the routine and wonder of human experience. Independent of the terms Rossi chose to describe it, the theater was a space "where the results of the most precise experiments [are] always unforeseen"—never unfolding precisely as scripted, never unfolding in precisely the same way across two performances.[8]

In concept, something of Rossi's vision of the theater-laboratory was realized a half-century prior to the publication of his observations. In 1930s Paris, architect Edouard Autant and actress Louise Lara conceived a theatrical work and its associated theater building as quite literally a theatrical machine (figure 2.1). Their *Théâtre de l'espace* performances were an empirical lens for examining "interpersonal relationships ... in real time, at real scale, and with real people"[9] that, in turn, generated the underlying concept for Autant's architectural design for theater building interiors. From Autant's careful study of the performances of each theatrical production, the architect drew inspiration for the design of a theater interior that would support that same social-spatial condition of the contemporary city that he and Lara reflected in the scripting and then observed in the performances. Within the physically constructed theater interior that resulted from this creative, generative process, the actors and spectators would subsequently "test" the supportive role played by the architectural design within the experimental realm of subsequent theatrical performances. Here, made visible and physical, was something akin to the iterative, cyclical process of design and evaluation familiar to HCI researchers. Here, too, was the visible, physical manifestation of Rossi's laboratory, the translation of theatrical performance through architectural design to the "bricks and mortar" of a building interior.

This give-and-take of the theatrical performance and the theater building, the interplay of spectacle and form, was of a kind Norbert Weiner anticipated in 1949 for computing: a "performance," he explained, determined by "impulses coming into it" as much as by a "closed and rigid set-up ... imposed ... from the beginning."[10] Indeed, the theater-laboratory of Autant and Lara exhibited many of the operative characteristics of computing displayed on display 3 considered in chapter 1: a *control system* whereby the particulars of the physical environment shape the interactions across people and things, and a *feedback system* whereby those interactions, in turn, shape the particulars of the physical environment.

Not surprisingly, architect Autant is exemplary of the architect-as-system-designer that Gordon Pask, a foundational figure of cybernetics, dated to the Victorian era. For Pask, architecture from this historical period forward was no longer a static, "classical" entity (characterized by formal purity, stability, and a legible style), but instead a dynamic system accommodating people and things in motion (characterized, in his words, by "development, communication and control").[11] In "The Architectural Relevance of Cybernetics" (1971), Pask identified *Parc Güell* in Barcelona (figure 2.2), the centerpiece of Antoni Gaudí's mostly unrealized design for a housing estate, as "one of the most cybernetic structures of existence."[12] Using the terms of systems theory, Pask observed that during one's visit to this surreal landscape, "statements are made in terms of releasers, your exploration is guided by specially contrived feedback, and variety (a surprise value) is introduced at appropriate points to make you explore."[13]

"Using physically static structures," wrote Pask, Gaudí had achieved a "mutualism" defined as "a dialogue between ... environment and its inhabitants."[14] What excited Pask was the prospect of expanding the mutualism achieved through more conventional (i.e., "static") architectural means by way of embedding computation in the very physical fabric of the architectural work. The outcome—a "reactive environment" for Pask—is a cyber-physical environment that, "within quite wide limits, is able to learn about and adapt to [its inhabitant's] behavioral pattern."[15]

In "The Architectural Relevance of Cybernetics," Pask identified Nicholas Negroponte and Christopher Alexander as among the very few young design researchers who demonstrated a "credible start" toward realizing this vision of a reactive, physical environment.[16] *A Pattern Language* (1977), arguably Alexander's most influential book, is essentially an elaboration of Pask's notion of mutualism: a cataloguing of the visual, spatial, and tactile information that is embedded in architectural systems of various scales and uses, and the capacity of this information to shape not only human behavior but, more broadly, Pask argued, the "conventions" and "traditions" of the culture of architecture and of society at large.[17]

Composed of vocabulary, syntax, and grammar, "each pattern" of *A Pattern Language*, Alexander explained, "represents our current best guess as to what arrangement of the physical environment will work to solve ... a problem which occurs over and over again in our environment."[18] For example, Pattern 185: Sitting Circle recognizes the problem of organizing a "group of chairs" in such a way as to "gather life around them."[19] In response to this problem, Pattern 185 (figure 2.3) mandates that each chair be placed "in a position which is protected, not cut by paths or movement, roughly circular, ... so that people naturally gravitate toward the chairs when they get into the mood to sit."[20] Alexander's intent was to suggest how the location and character of various physical elements of the environment (their physical pattern) shape the interactions across people and things (their behavioral patterns).

While in *A Pattern Language*, Alexander exhaustively elaborates the workings of environment-inhabitant mutualism, it is only in the final vignette, Pattern

2.2
Parc Güell (Antoni Gaudí, Barcelona, 1900–1914), "one of the most cybernetic structures in existence" (photograph by the author).

2.3
Pattern 185: Sitting Circle (from Christopher Alexander's *A Pattern Language* [1977], reprinted by permission of Oxford University Press, USA).

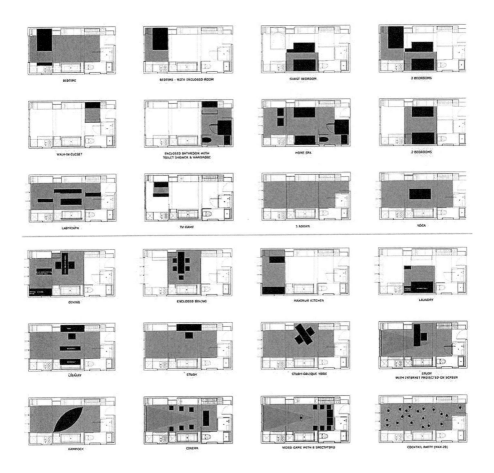

2.4
Gary Chang's Domestic Transformer: twenty-four configuration patterns (diagram by Gary Chang).

253: Things From Your Life, that we come to a room exhibiting the kind of vitality Pask envisioned: a room filled with "things from your life" that for Alexander comes "gradually to be a living thing."[21] This vitality is reliant more on human perception than on anything having to do with the physical room itself, a room that is no more than a conventional, "physically static" container for personal effects. While human perception of and ascription of agency to our living environments is an important, necessary, and difficult consideration of *this* book, in Alexander's book, the room remains just a room to the end: a room not yet embedded with anything of computing, a room not yet affording any alterations to its spatial form, texture, or color other than by the costly, laborious means of conventional building construction (e.g., by constructing or demolishing walls or by painting surfaces).

Nevertheless, *A Pattern Language* has notably affected computer science, among other fields.[22] For humanoid robotics research in particular, *A Pattern Language* has provided the conceptual model for choreographing human-robot interaction.[23] Specifically for architectural robotics, *A Pattern Language* offers the concept of pattern making as a means for defining the physical configurations assumable by a reconfigurable environment in its support of human activities.

Moreover, *A Pattern Language* offers architectural robotics a foundation for an adaptable, reconfigurable environment, as anticipated in Alexander's idea of "compressing" two or more patterns into a single space. To illustrate this compression of patterns, Alexander envisioned all of the functions of a typical house occurring in the space of a single room,[24] resulting in a building that, in practical terms, exhibits an "economy of space" that is potentially "cheaper" to realize. A recent demonstration of Alexander's "compressed" house is the Domestic Transformer, a 330-square-foot, single-room home in Hong Kong of sliding walls and hinged panels, manually reconfigured by its owner-architect, Gary Chang, to fashion any one of twenty-four different living patterns (figure 2.4).[25] But while Chang's home is compelling and informative for architectural robotics, it is not robotic, interactive, or intelligent. And for Alexander, beyond flexibility, compactness, and potential cost savings (as relevant to this Hong Kong apartment), a compressed home should be fundamentally "poetic," offering in its compacted, patterned layers a "denser" meaning for its inhabitants.[26] As Alexander maintained, "this compression of patterns illuminates each of the patterns, sheds light on its meaning; and also illuminates our lives, as we understand a little more about the connections of our inner needs."[27] (The poetry of manually reconfigurable homes, found in the traditions of Japan, and conceived more contemporaneously by Gerrit Rietveld and Carlo Mollino, is considered in the next chapter.)

Overall, *A Pattern Language* lends to the emerging field of architectural robotics the precedent of a carefully conceived physical environment affording intimate and evolving relationships across people, things, and physical space.[28] Architectural robotics aims to afford the heterogeneous relationships of human and non-human actors, as anticipated by Alexander's *A Pattern Language*, and also characterized more recently by the "Internet of Things," as well as by the "material-semiotics" of actor-network theory (ANT) in which people and things coexist as actors, as agents, in a dynamic interplay of the animate and inanimate.[29]

Of the two younger figures identified by Pask in "The Architectural Relevance of Cybernetics," it was Nicholas Negroponte, associated recently and visibly with the XO (One Laptop per Child) Project, who assumed the formative role in realizing the early promise of intelligent environments. Drawing on the work of his Architecture Machine Group (which evolved into the MIT Media Lab),[30] Negroponte, in *Soft Architecture Machines* (1975), echoed Pask's concept of a "reactive environment." For Negroponte, a "reactive environment" (which he later, in *Soft Architecture Machines*, refers to as a "responsive environment," and more consistently as an "intelligent environment") is distinguished from the "simple sensing-effecting model of computation," in which "a processor receives signals from its sensors and emits responses with its actuators."[31] An example of the latter is the strictly "computer-controlled conventional homes" regulating temperature and artificial lighting and furnished, as Negroponte wryly offers, with a "sofa that alters itself to 'fit' the body aloft, and that initiates soporific music and smells at 10:30 PM."[32] Given the ubiquity of sensors, actuators, and microcontrollers today—their affordability and ease of use—we are all too familiar with the kind of interactive works that Negroponte largely dismissed forty years ago. This kind of computational artifact, Negroponte argued, suffers from the simplistic "thermostat analogy," where heating is mechanically delivered to the building interior whenever the temperature dips below the set temperature. Instead of the simplistic, hard-wired response system exhibited by the thermostat, Negroponte and his group envisioned something more ambitious for the meeting place of computing and the built environment: "artificial domestic systems capable of intelligent responses."[33]

Negroponte devotes chapter 4 of *Soft Architecture Machines* to the characterization of such "intelligent environments," summarized as follows:

- Intelligent environments have sensors and actuators.[34]
- Intelligent environments are "responsive to you and me and the world."[35]
- Intelligent environments "map the occupants' activity and observe the results."[36]

- Intelligent environments "can grow and upgrade themselves."[37]
- The responses of intelligent environments are not predetermined—"some 'learning' is involved."[38]
- The "behavior [of intelligent environments] is purposeful if not [strictly] intelligent."[39]

If the last three of these six listed items are any indication, Negroponte is not upending the thermostat analogy absolutely, but rather by degree. With respect to the "intelligence" of an intelligent environment, Negroponte qualifies that intelligent environments exhibit *some* degree of machine learning and adaptation and exhibit machine intelligence *if and to the extent that* intelligence serves a purpose. Consequently, a segment of the artificial intelligence (aka AI) research community probably views intelligent environments as not sufficiently "intelligent" by classical AI standards. Nevertheless, the *intelligence* of intelligent environments assumes, for Negroponte, "an active role, initiating ... changes as a result and function of complex and simple computations."[40] More profoundly, intelligent environments promise to foster "a dramatically different relationship between ourselves and [them]" by configuring meaningful places for human experience.[41]

An anticipated extension of Negroponte's vision of intelligent environments was, for William Mitchell, buildings "that become ... more like us. We will continually interact with them, and increasingly think of them as *robots for living in*."[42] Mitchell's notion of "robots for living in" perceptibly anticipates the emerging field of architectural robotics, an unsurprising extension of intelligent environments defined by the movement of not only *bytes* but also *bits* in the physical sense (i.e., mass). Whereas Mitchell expressively cites *buildings* as "robots for living in," architectural robotics is more inclusive with respect to physical scale, so that its artifacts—the products, manifestations, or things of architectural robotics—begin with the scale of furniture and extend to the scale of urban infrastructure and into the metropolis. More than its precise physical measurements, it is important to recognize that the artifacts of architectural robotics are relatively large—*larger* than the physical scale of artifacts of tangible computing and industrial design. This means that the architectural robotic artifact is of a scale that challenges a human being's capacity to objectify it. In other words, the artifacts of architectural robotics are either spatial (they have, minimally, two walls in reasonable proximity to one another that together establish an enclosure or spatial frame) or compete in size with our physical bodies (as does a dining room table, and certainly as does a suite of dining room furnishings). Accordingly, architectural robotics, given its imposing physicality, represents a different kind of human-computer and human-robot interaction. We cannot objectify such artifacts as we can, say, a mobile phone that fits in our hand or our pockets. We engage architectural robotic artifacts in a state of distraction, as they surround us or at least challenge the physical scale of our bodies. For this reason, the Viennese architect Adolf Loos discouraged the photographing of his architectural works: How could a photograph possibly capture their essence when that essence escapes any and every vantage?

The physicality, behavior, and affordances of architectural robotics have the potential to make architectural robotics an intimate aspect of our private and social routines. In the near future, in our everyday lives, likely places in which we will encounter architectural robotics include health-care facilities, classrooms, workspaces, assisted-care homes, and within mass public transit systems. Applied to these familiar building typologies, architectural robotics has the potential to render these common places more responsive, supportive, and pleasurable (or, at least, less painful).

Outside the course of such daily routines, architectural robotics may prove invaluable in instances of disaster relief, in making the physical environment more agile in response to natural and human-made disasters. And with increasing mass urbanization, architectural robotics promises to transform underutilized spaces of single-program buildings, such as schools, sports arenas, and offices, into the dense, compressed-pattern programming that Alexander envisioned for the home, enabling living, working, and recreation opportunities to occur within the same spatial, physical envelope in places where space is too scare and/or exceedingly costly to devote to single purposes that are not used continually.

Given the foreseeable reach of architectural robotics and the intimate role it may assume in our lives, the design of such large-scale, cyber-physical artifacts demands a deep understanding of human populations sometimes overlooked by designers: children, adults aging in place, and individuals of all ages suffering from illness, disabilities, and the challenges of dislocation and relocation. It is particularly in this respect—the understanding of human beings and being human—that architectural robotics research benefits from engagement by human factors psychologists and allied cognitive and social scientists. The design of architectural robotic artifacts also demands that research teams capably apply this understanding of human populations to the conception of artifacts that are at once digital, physical, spatial, and relatively large scale. Such complex artifacts must not only be easy to use but also safe to use. They must be durable, cleanable, robust in function, and aesthetically and functionally pleasing to their users. Both at rest and in motion, they must be able to support their own weight, carry loads placed on them, and maintain their balance. In short, designing architectural robotic artifacts demands an understanding of the needs and wants of human users engaged in routine activities and episodic events, as well as the broad and deep knowledge required for the development of responsive, larger-scaled, cyber-physical artifacts that will operate in close proximity to human beings and be characterized by the movement of physical mass (those *bits*).

The complex interdependencies of architectural robotics—between the designed cyber-physical artifact and the human needs and wants it supports—typify a research problem characterized as "underconstrained." Underconstrained problems (also referred to as "underdetermined problems") are those for which no optimal solution exists. Underconstrained problems are endemic to the real-life applications of architecture, architectural robotics, and broadly HCI, where human needs and wants, together with the artifact's specifications addressing these, cannot be adequately characterized because of numerous, onerous, and often conflicting variables, objectives, and/or stakeholder interests. Given their unwieldy nature, underconstrained problems have been described by Rittel and Webber as "wicked": wicked, because such problems are often the symptoms of still other problems, and their "solutions" are essentially the result of trial-and-error processes (however we strain to control them) that cease without an ultimate test of validity, and mostly for reasons external to the problem (e.g., constraints of time and money).[43]

Writing specifically from the perspective of HCI design-researchers, Zimmerman, Forlizzi, and Evenson recognize in wicked problems an opportunity not only for scientists and engineers but also for *creative* designers.[44] This is certainly the case for architectural robotics research, which, by definition, is the hybrid of two paradigms: architecture (and its allied design fields) and robotics (and its allied computing and engineering fields). Given that "design," as noun and verb, implies a seemingly unbounded range of artifacts and actions, care must be taken, indeed, with any terminologies containing it, starting with the word *design* itself, by which we mean the *creative* design activity performed by the likes of architects, graphic designers, and industrial designers, compared to the design activities performed by, for instance, software developers and mechanical engineers intended either for commercialization and/or to meet an explicit set of specifications. By "design research," we mean a research endeavor aiming to *literally* produce knowledge: to create artifacts through a careful process of making (i.e., designing, prototyping, and testing) that is itself a contribution to knowledge, manifested as "design precedents" (for architects) or as "design exemplars" (for HCI researchers).[45] The iterative design process is, in turn, informed by "design thinking," which involves, to borrow terms from Zimmerman and colleagues, *grounding* (an understanding of the problem, gained by drawing on multiple perspectives), *ideation* (the generation and evaluation of alternative design solutions), *iteration* (the iterative process of "refining a concept with increasingly fidelity"), and *reflection* (a meticulous documentation of the design process and an evaluation of its outcomes "in a way that the community can leverage the knowledge derived from the work").[46]

Following from the consideration of wicked problems by Zimmerman and colleagues and the larger HCI community, it is the "wickedness" of real-life applications of architectural robotics that brings together creative design rooted in architecture, and robotics rooted in engineering and science. Too frequently, human problems and opportunities have been addressed from an engineering and science perspective seeking a "technical solution." While engineers and scientists play indispensable roles in responding to underconstrained human problems, we wish to emphasize, specifically for architectural robotics, the "opportunity for design researchers," as recognized by Zimmerman, Forlizzi, and Evenson, "to provide complementary knowledge to the contributions made by scientists and engineers through methods unique to design and design processes."[47] Indeed, architectural practice has been defined for millennia by the integration of *art* and *technology*—the two buzzwords used to describe a new flavor of interdisciplinary academic programs offered by research universities that also form, for primary and secondary education, the current revision of STEM as STEAM (science, technology, engineering, art, and math).

It is the collaboration of engineers, scientists, and creative designers of various kinds that arguably offers the most promising conditions for an architectural robotic response to the most wicked problems of living today. In this collaboration, however, it is the role of the creative designer that is most unclear and warrants more clarification. In "Designing as a Part of Research," Pieter Jan Stappers inventories the virtues designers bring to research pursuits that, I would argue, extends to architectural robotics. These include, for Stappers, "the ability to integrate findings from different disciplines; to communicate with experts of different disciplines; to keep in mind the interests of all different stakeholders (e.g., user, technology, business); to take decisions and make progress in the light of incomplete information; to maintain a focus on the aim."[48]

"The aim," Stappers contends, is the "prototype," which serves "as a kind of working hypothesis, not necessarily a static one that is tested and refuted or proven to be 'true'; but possibly a dynamic one that is adjusted, grown, and shown to work."[49] Striving to address the wicked problems of living today, the artifacts of architectural robotics—prototypes viewed as working hypotheses as much as solutions—benefit from tethering the line between art and science, architecture and engineering. Consequently, the promise of architectural robotics is not the isolated object but instead a cyber-physical environment that supports and augments an increasingly digital society. An architectural robotic environment that cultivates interactions across people and things has the potential to cultivate places of social, cultural, and psychological significance within our complicated, complex, and wondrous world.

THIS BOOK

This book focuses on architectural robotics of modest, physical scale; that is, cyber-physical artifacts larger than those typical of tangible computing, and smaller than the outcomes typical of architectural practice. The consideration of architectural robotics is illuminated by stories and metaphors: tropes that help communicate the essence and reach of architectural robotics. The epilogue of this book considers, finally, an architectural robotic "future," scaled-up as cyber-physical neighborhoods, villages, and metropolises.

Specifically, this book focuses on architectural robotics at the physical scale of furniture, workstations, and rooms. Three research projects are elaborated in detail in three book parts devoted to them—the efforts of my research lab at Clemson University. Contextualized in a variety of ways, these three cases are early *design-research exemplars* of three typologies for architectural robotics, each of which defines a chapter of this book and also anchors one of its three parts. In addition to the three case studies, the book considers the corresponding motivations, theories, interaction modalities, and technologies that further define the three typologies, which are introduced in the following three subsections.

The Reconfigurable Environment ("Makes Room for Many Rooms")

Essentially Alexander's concept of "compressed patterns," where all of the functions of a typical house occur in the space of a single room, the reconfigurable environment is a malleable, adaptive environment specifically dependent on moving physical mass to arrive at its shape-shifting functional states. The case study for this part is the Animated Work Environment, a robotic work environment that dynamically shapes and supports the working life of multiple people, collaborating together in a single physical space, who use digital and analog tools in the creation of digital and physical artifacts. The foci of this chapter are (with respect to technology) kinematics and (with respect to human-machine interaction) cyber-physical configuration patterns conducive to working life in an increasingly digital workforce. *The remarkable feature of the reconfigurable environment typology is that a single physical space makes room for multiple functional places supporting and extending human activity: "one space, many rooms."*

The Distributed Environment (Where "Furnishings Come to Life")
Not unlike a distributed computer environment, the distributed environment typology is defined by a collection of cyber-physical devices—each having different functionalities tuned for different purposes—that are networked together and recognize and adapt to one another and to their users to support a common human need or to exploit an opportunity to improve human life. The case study for this part is *home+*, a suite of networked, distributed "robotic furnishings" integrated into existing domestic environments (for aging in place) and health-care facilities (for clinical care). The foci of this part are (with respect to technology) compliant, "soft" robotics; a nonverbal, gestural form of human-machine interaction based in machine learning; and (with respect to human-machine interaction) designing specifically for vulnerable human populations. *The remarkable feature of the distributed environment typology is that, seemingly, the familiar furnishings found at home, at work, at school, and within other everyday places "come to life," forming a "living" room.*

The Transfigurable Environment ("a Portal to Elsewhere")
Like the reconfigurable environment, the transfigurable environment is a malleable, adaptive environment dependent on moving physical mass. In the transfigurable environment, however, the changes in state are not principally for the purposes of providing different functional programs matched to changing human needs and wants, but for the purposes of transporting the inhabitant, seemingly, to some other place, perhaps far away. The transfigurable environment is in this sense what Negroponte defined as a "simulated environment" where "one can," for example, "imagine a living room that can simulate beaches and mountains."[50] The case study for this part is the LIT ROOM, a robotic room that is embedded in the everyday physical space of a library and is transformed by words read by its young visitors so that the everyday space of the library "merges" with the imaginary space of the book. *The book is a room; the room is a book.* The focus of this chapter is the iterative codesign process (which includes children and teachers) to develop a cyber-physical artifact conducive to collaborative learning activities. *The remarkable feature of the transfigurable environment typology is that a single physical space becomes "a portal to elsewhere."*

STORIES AS DESCRIPTIVE, PROPHETIC INSTRUMENTS
Metaphor has already proven to be a powerful starting point for this book, both in defining milestones of HCI and in locating architectural robotics at the interface of architecture, computing, and the theatrics of everyday life. As this book is very much about the enterprise of building architectural works with embedded

computing, Gaston Bachelard's elegant characterization of metaphor as the "working draft" from which to "build a house" seems particularly befitting.[51]

More than metaphor, however, it is "storytelling" that will be especially productive toward capturing architectural robotics in the chapters of this book that follow. Weiner's prophetic essay of 1949, "The Machine Age," is, once again, remarkable in situating the importance of storytelling in describing complex constructs such as architectural robotics. As Weiner remarked, "In the discussion of the relation between man and powerful agencies controlled by man, the gnomic wisdom of the folk tales has a value far beyond the books of our sociologists."[52] Similar to the sentiment of Weiner, but from the vantage of someone working actively today at the intersection of computation and storytelling, is the offering of Pixar's Andrew Stanton: that storytelling has the capacity to uncover "truths that deepen our understanding of who we are as human beings."[53] And at the core of her own storytelling "about imagined futures," Margaret Atwood recognizes her preoccupation with fundamental questions on "human nature": "Who are we, really? What are we capable of? How far can we change without ceasing to be ourselves?"[54]

Within the sphere of HCI itself—at CHI, its premier international conference, to be precise—storytelling and the metaphysical questions it conjures were the focus of a conference workshop entitled "Alternate Endings: Using Fiction to Explore Design Futures."[55] Participants in this workshop were asked to use "fictional narratives to envision ... the values and implications of interactive technology."[56] The directive from the workshop conveners follows from Nathan et al., who propose "the use of short, speculative narratives, or value scenarios, as a method of inspiring critical reflection" on contemporary HCI projects.[57] It is precisely this procedure, a reliance on storytelling to frame not only the design of architectural robotics, but also its "social, psychological and ethical dimensions," that we will employ here to frame each chapter.[58] Dwelling in stories may prove revelatory: as we come to live, play, and work increasingly with architectural robotics, our world will become much more like the stories we tell today.

RECONFIGURABLE

3 SPACES OF MANY FUNCTIONS

A house in construction symbolizes our burning passion for the
coming-into-being of things. Things already built and finished,
bivouacs of cowardice and sleep, disgust us! We love the immense,
mobile, and impassioned framework that we can consolidate,
always differently, at every moment....

—F. T. Marinetti, "The Birth of a Futurist Aesthetic"

Imagine a workplace that physically reconfigures itself to support our changing needs. Introduced in the first chapter of this book, the Animated Work Environment (AWE) is one example: a workstation composed of multiple work surfaces and a robotic wall, which is faced with computer and analog display ribbons (see figure 3.1, the physical realization of the digital visualizations of figure 1.3). Now imagine that someone seated at AWE, working intensively for two hours, suddenly stands up to welcome a close collaborator who has just entered the workspace. AWE detects the abrupt movement and responds, pulling its robotic ribbon-wall upward and out of the pathway of the departing worker. Or imagine that someone seated at AWE swiftly transfers attention from a journal article taking shape across AWE's multiple display ribbons to the more manual, physical task of rummaging through stacks of printed scientific journals to locate a reference. AWE shape-shifts to accommodate the change in work state, configuring itself for working with printed documents. Or then again, maybe someone seated at AWE simply has had enough of working intensively on that journal article, but nevertheless has to trudge onward to meet the submission deadline. Might AWE detect this change in spirit and alter its lighting and physical (spatial) configurations accordingly to help make the activity less arduous? As will be elaborated in this chapter, AWE responds to changes in the workflow, recognizing the need for reconfiguration that will better support it. AWE physically reconfigures, partly by its users manually adjusting AWE's physical components, partly by users gesturing toward AWE's infrared (IR) sensors, and partly by AWE responding to a change in the activity or behavior of its users. The development of AWE required the design team to consider all these pathways to reconfiguration: to recognize in one form still other forms.

This is a very different way to think about form for designers and (especially) architects who practice their vocation under the assumption that form is singular and stable. Art historian Henri Focillon thought otherwise, grappling with the notion that a single form is not singular or stable, but rather has within it a multitude of forms. "Although form is our most strict definition of space," wrote Focillon, "it also suggests to us the existence of other forms."[1] The eclectic Jesuit theologian and scholar Michel de Certeau asserted much the same, but

3.1

The Animated Work Environment (AWE), as it appeared in the January-February 2012 issue of *Interactions* magazine (reprinted with permission of the author and ACM *Interactions*).

more specifically for the built environment: that a place is more "a piling up of heterogeneous places" than a singular entity, despite the authoritative stance of architecture and engineering in giving definite form to them.[2] Countering this position, Focillon stresses that we must "never think of forms, in their different states, as simply suspended in some remote, abstract zone.... They mingle with life, whence they come; they translate into space certain movements of the mind."[3] As we conceive them and live with them and *within* them, "each form," writes Focillon, "is in continual movement, deep within the maze of tests and trials."[4]

We find suggestions of this kind of reconfigurability in some corners of our everyday environment. If we consider reconfigurability as occurring over many years, an obvious example of reconfigurability in the built environment is the medieval church, which, in many instances, is a composite of often dramatically different building fabrics grafted onto the previously existing structure.[5] Reconfigurability can also occur, not over decades or years, but in real time, in for instance the mattresses commonly found in hospital beds that reconfigure in form to prevent bedsores. But recognize that the reconfiguration of a hospital mattress occurs at programmed time intervals and not at precisely the instant when a particular patient most requires it. The reconfiguration of the hospital mattress exemplifies *automation*—in this case, a process happening at intervals timed to prevent bedsores in the typical user. Consequently, the reconfiguring mattress does not exemplify *interactivity* or *intelligence*, where the user's actions drive the configuration change. More interactive and at least party intelligent, architectural robotics, unlike the automated mattress, aims to be very attuned to users' needs in real time.

Another suggestion of reconfigurability—more sophisticated than the mattress example—comes as likely equipment for cars of the near future that senses a driver verging on sleep and, in response, provides a vibration to the steering wheel—a physical jolt to the driver's awareness.[6] While this response follows not from a programmed interval (in the mattress case) but originates in the user (nodding off), the response comes in the form of a jolt rather than, say, changes in interior lighting, sound, temperature, air circulation, the form of the seat and/or the cockpit—the more subtle, ambient modifications to the driving experience that might encourage the driver to carry on, maybe even pleasurably, to the intended destination. This kind of interactive and intelligent, ambient modification characterizes the ambition of architectural robotics: to sense conditions external to it, to render appropriate decisions, to alter the environment accordingly (particularly, its physical form), and possibly to learn from its interactions with inhabitants how it might perform in the future in a more suitable manner.

FORM AND FLUX

With few exceptions, designing the built environment for this kind of reconfigurability has been resisted by the discipline of architecture throughout human history. The resistance from architecture has a functional dimension as well as an aesthetic, formal dimension.

From the functional side, resistance to reconfigurability is motivated by the requirement of architecture to protect its inhabitants from the elements and by the desire for it to maintain the everyday life practices and customs of those inhabitants. In short, architecture is the physical embodiment of our human need to resist natural and societal threats: the rage of storms and the entry of the unexpected, the unwelcomed, and the unknown. In this light, architectural works, in and of themselves, must not change, as we depend on them to maintain continuity, to defy or at least to resist the impositions of nature and unfamiliar humankind. Consequently, we spend a great deal of time and money maintaining and restoring architectural works in order to maintain their performance and the illusion of sociocultural and political stability they edify.

Today, resistance to reconfigurability is complicated because of a combination of functional, economic, legal, and frequently political demands imposed on architectural works that has closed the minds of clients and of many architects to allowing room within the architectural work for functional play. Architectural works are not typically multifunctional but typological: they accommodate a narrow band of human activity. This is why our schools and office buildings are mostly empty overnight, our homes are mostly empty during the day, and our theaters and arenas are empty most of the time. What a waste of resources, at the expense of a growing segment of the human population seeking basic shelter or affordable places to meet, work, and play. Conventional homes and workplaces, meanwhile, are incapable of responding to changes within individual inhabitants as they grow, grow old, and sometimes grow sick and to groups that grow and shrink in their numbers and that have varied and fluctuating needs. The discipline of architecture has mostly ignored the flux endemic to life, perpetuating instead monuments of various and sometimes exuberant expression that may delight us but most often maintain a sense that nothing changes. Architecture has been, historically, petrified of change.[7]

From the aesthetic, formal side of architecture, resistance to reconfigurability is motivated by the quest for a universal standard for measuring it: for designing its component parts and organizing these parts to constitute the larger work. From Vitruvius to Le Corbusier—the better part of two millennia between them—the dimensional and proportional systems of architecture were modeled on an idealized and yet motionless human body: the Vitruvian and Modular men (figure 3.2). Maintaining the continuity bridging these two figures, Alberti

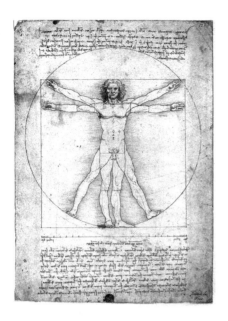
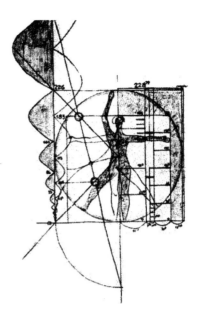
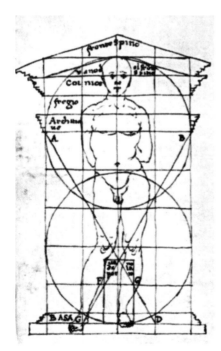

3.2

The Vitruvian and Modular men of architecture, conceived by Vitruvius (ca. 27 B.C.), Le Corbusier (1955), and Francesco di Giorgio (ca. 1482). **(Top left)** Vitruvian Man, as drawn by Leonardo da Vinci in 1492 (source: Wikimedia Commons). **(Top right)** Modular 2, as drawn by Le Corbusier (reprinted with permission, © F.L.C. / ADAGP, Paris / Artists Rights Society [ARS], New York, 2014). **(Bottom)** Vitruvian Man, as drawn by Francesco di Giorgio (reprinted with permission of MIT Press).

3.3
A traditional Japanese house interior
with sliding shoji screens
(source: Wikimedia Commons).

and his peers in the Renaissance maintained the classical rule: "the reasoned harmony of all the parts within the body" of an architectural work, with each of its parts being designed and constructed following from the established rules of its particular "Order" (e.g., Dorian, Ionic, or Corinthian, as expressed most visibly in a building's columns).[8]

Also from an aesthetic point of view, the discipline of architecture has rejected the capacity for reconfiguration when it maintains, as it often still does, the Renaissance ideal of a "beautiful" architecture, exemplified by architectural works for which "nothing may be added, taken away or altered."[9] A beautiful architecture is one that cannot be modified and, also for Alberti, one that is timeless ("classical"). Even well into the twentieth century, the discipline of architecture stubbornly advocated the timelessness of architectural works, evidenced by, among others, the Italian Rationalists, Hitchcock and Johnson and their "International Style," and Le Corbusier and his intensive post-war efforts to establish his Modular as the universal measure for a beautiful, modern architecture.

This "immobility" of architecture has its historical exceptions. Almost all buildings have functioning doors and windows, and it is not entirely novel for a piece of furniture or even a building interior to permit changes to its physical form to afford different functions supporting different objectives or activities. Sliding partitions in conference centers divide meeting rooms to accommodate different conference sessions tailored to the expected number of attendees. Retractable stairs divide the interior of a home into living space and attic. Hinged, retractable surfaces transform the lectern seats within a university auditorium into writing desks. These kinds of mechanical affordances or *action possibilities* date back centuries, for example, in the form of tatami mats and sliding shoji screens found in traditional Japanese houses (figure 3.3).[10]

Most notably in architecture, the Rietveld Schröder House (1924, Utrecht), designed by cabinetmaker and architect Gerrit Thomas Rietveld, extended the concept of the sliding screen to permit the manual reconfiguration and repositioning of various components of the home's second story. Wall partitions slide to close off living spaces or to open them up, almost entirely, to make a continuous, second-floor living environment (figure 3.4). Unornamented shutters mounted to the walls are repositioned to close off or open large expanses of ribbon windows to (what was originally) an extended prairie. Hinged panels open and close off the bathroom and stairwell, and also the skylight above, accessed by a specially designed step ladder. Likewise on the stairwell landing, a sliding panel controls access from the first floor to this wondrous, animated second level.

Carlo Mollino, a mid-twentieth-century architect who was known for his own reconfigurable architectural contrivances, imaginatively characterized the

3.4
Second-level interior of the Rietveld Schröder House (Gerrit Rietveld, Utrecht, 1924): **(Top)** closed and **(Bottom)** opened (reprinted with permission of the Rietveld Schröder Archief, Centraal Museum Utrecht, Bertus Mulder/ Pictoright, © 2014 Artists Rights Society [ARS], New York).

Chapter 3

manually reconfigurable house as "a jack-in-the-box, a play of easily changeable rooms and furnishings, a fickle scenography of embroidered furnishings and sliding, transforming rooms, separating and creating halls and lounges with the turn of the seasons, in states of animation, reflecting the ceremonies of 'domestic' happenings.... When importune, the furnishings truly disappear into the wall."[11]

The "easily changeable rooms and furnishings" Mollino describes are alive with possibilities for reconfiguring them. What fascinated the Turinese architect was not so much the physical movement afforded by the sliding partitions and furnishings (their mechanics), but mostly how these architectural elements, in their flexibility, reflected things external to them: the passing of the seasons, the unfolding rituals of domestic life. In his own architectural works, Mollino invited the inhabitants of his rooms and furnishings to tune their mechanical features to reflect the conditions of their interior lives—to reflect themselves in the environments in which they live, *to make themselves more at home*. In the number of interior domiciles he both designed and took refuge in, the unpredictable Mollino so much sought the same restfulness for himself, but he recognized, in states of torment and elation, the difficulty of capturing this state of peace, even for the duration of the shutter movement of his Leica camera (figure 3.5).

3.5
Carlo Mollino, capturing himself by camera in the interior of the *Casa Miller*, which Mollino designed and owned (reprinted with permission of the Politecnico di Torino, Sistema Bibliotecario, Biblioteca Centrale di Architettura, Archivi, Fondo Carlo Mollino).

Despite the best efforts of the discipline, architectural works are no more static than the lives living within them. When we enter a building, we bring with us the dimension of time. No inhabitant will ever have precisely the same experience here, nor will any other inhabitant have precisely the same experience here as someone else inhabiting the same space. The human experience, framed by the physical environment, is never precisely the same at two points in time. For this reason, the poet F. T. Marinetti, founder of Italian Futurism, expounded the virtues of the house "under construction." The framework, the scaffolding of architecture, suggested to Marinetti the possibilities of what it may become, a work incomplete but promising, *the coming-into-being of things*, like ourselves. Because the building under construction reflects our own aspirations-to-become, "we love [its] immense, mobile, and impassioned framework that we can consolidate, always differently, at every moment."

It should be quite apparent that an architectural work that is reconfigurable is one that, at least in conventional architectural terms, is *unfinished*: room is made, in the very design of such a place, for the inhabitants to, in a word, play. Architectural works, like all works of art, are "quite literally 'unfinished,'" argued Umberto Eco: "the author seems to hand them on to the performer more or less like the components of a construction kit."[12] For Eco, as might be said for Marinetti, "the comprehension and interpretation of a form can be achieved only ... by repossessing the form in movement and not in static contemplation."[13] Implicit in Eco's remark is the notion that the artifact's vitality, its liveliness, opens it to our comprehension. Why is this so? Why is it that the vitality of things, their flux, lends understanding to the things we experience? Because "flux is reality itself," the "consistency" in our life experience.[14] In this book, I argue that certain aspects of the built environment will make most sense to us, will support us and expand our opportunities most, when they actively grow and adapt with us over time.

A PROTEAN ARCHITECTURE FOR A PROTEAN PEOPLE

The means of computing and, in particular, robotics can be integrated into the physical fabric of architectural works to forge a more interactive, more intimate relationship between the built environment and us. Embedding these digital technologies in selected aspects of the larger built environment renders them a semblance of vitality: the capacity to move with and respond to things external to them, whether these things are living (people and their pets), or inanimate (physical property), or far less tangible (data streaming over the Internet, the detection of weather). But whatever the source (the input), for the purpose of this book the vitality exhibited by the built environment is defined primarily by its physical shape-shifting that results from the unfolding of the following process:

1. its sensing some change (typically, motion) in the bits, bytes, and biological systems inside and/or around it;
2. its decision making based on this input; and
3. its corresponding actuation of physical mass(es) within it: this, accompanied by some degree of machine intelligence by which it remembers, recalls, and/or recalibrates its responses.

In this very active way—of engaging the world, drawing inferences, and responding in kind—the architectural robotic environment is to some degree a reflection of us: our needs, our aspirations as vital beings.

Shape-shifting has fascinated us for millennia, as evidenced by its centrality in classical Greek mythology. As Steven Levy asserts in *Artificial Life*, today's human-made, life-like artifacts are founded not only in the contemporary imagination, but equally so in the many "ancient legends and tales" devoted to the theme of "inanimate objects" infused "with the breath of life."[15] Following Levy and the a-life community and recognizing physical reconfigurability as a pathway, today—to a more intimate correspondence between our built environments and ourselves—it is not such a stretch briefly to consider the myth of Proteus (figure 3.6), the Greek god who, more than other shape-shifters in Greek mythology, was capable of transforming himself into countless different forms. This captivating capacity of Proteus to shape-shift led to his ultimate "transformation": into the familiar adjective, *protean*, which wonderfully captures the core behavior of architectural robotics.

Proteus assuredly enters our cultural imagination in *The Odyssey*.[16] In this epic work by Homer, the shape-shifting Proteus, god of the sea, willfully changes form like the ebb and flow of the tides. In Homer's telling of the tale, Proteus holds invaluable secrets about navigating the fierce seas that would aid Menelaus, the stranded warrior, in his homeward voyage by ship. Proteus' daughter, charmed by Menelaus, offers the pitiful warrior guidance on how best to obtain the secrets from her father:

> He'll attempt / to change himself into all sorts of shapes / of everything that crawls over the earth, / or into water or a sacred flame. / You must not flinch —keep up your grip on him.
> **(Book IV, 414–418)**

As the story continues, Menelaus seizes hold of Proteus, and Proteus eludes the warrior by changing form:

> He turned himself into a lion, / and then into a serpent and a leopard, / then a huge wild boar. He changed himself / to flowing water and towering tree.
> **(Book IV, 445–458)**

3.6

Proteus, the shape-shifter: an etching from *Der höllische Proteus* by Erasmus Francisci, 1695 (source: Wikimedia Commons).

The struggle between Menelaus and Proteus unfolds just as the daughter described it: Proteus, an old man at the core, eventually tires of shape-shifting and yields his secrets of the sea to Menelaus.

Despite his advanced age and waning stamina, the Proteus of this poem from the eighth century B.C. has led an active and prolonged life, under the same name but in different guises.[17] Notably, Proteus is the name given to characters in Milton's *Paradise Lost* and in Shakespeare's *Henry VI* and *The Two Gentlemen of Verona*. Proteus is also the name given to historic warships (both USS *Proteus* and HMS *Proteus* of the Royal Navy) and to a novel contemporary vessel (the Proteus WAM-V, which features a reconfigurable hull that conforms to the surface geometry of water currents). Proteus is also the name given to, respectively, a medical syndrome popularly identified with "the Elephant Man," a bacteria having a remarkable ability to evade the host's immune system, and a family of flower having more than 1,400 varieties. Our fascination with shape-shifting is evidenced not only by this extended and variegated procession of forms under the name Proteus, but also by the contemporary usage of the word *protean* defined as *adopting or existing in various shapes, variable in form; able to do many different things;* and *versatile*.[18] All of these definitions aptly describe architectural robotics.

There remains one more Proteus, however, that will prove useful in uncovering the promise of the particular kind of human-computer interaction that architectural robotics affords. This is the Proteus of psychology. Both the Proteus of Heinrich Khunrath, the sixteenth century German physician-alchemist, and the Proteus of Swiss psychologist Carl Jung in the twentieth century personified the elusive unconscious. But for our purposes, the more useful Proteus is the one that names a contemporary, psychological profile, as considered by psychiatrist Robert Jay Lifton. In *The Protean Self: Human Resilience in an Age of Fragmentation*, Lifton characterizes this modern-day Proteus as "fluid and many sided" and "evolving from a sense of self [that is] appropriate to the restlessness and flux of our time."[19] This Proteus, a "willful eclectic," draws strength from the variety, disorderliness, and general acceleration of historical change and upheaval. As Lifton writes, "One's loss of a sense of place or location, of home—psychological, ethical, and sometime geographical as well—can initiate searches for new 'places' in which to exist and function. The protean pattern becomes a quest for 'relocation.'"[20] According to Lifton, the protean self actively responds to life's challenges and opportunities—whether pedestrian (working life, family life) or grand-scaled (social, economic, political)—by seeking "places" best suited for improvement, advancement, or at least escape.

The protean way—to be fluid, resilient, and on the move—is not only a tactical, cognitive response to living today, but is, according to anthropology researchers Antón, Potts, and Aiello, *the outstanding trait distinguishing the human species*. The protean way is defined as "adaptive flexibility," the cornerstone of this new paradigm for human evolution, as published by the three researchers in the journal *Science*. Antón, Potts, and Aiello find evidence for adaptive flexibility in all the "benchmarks" defining our species: "dietary, developmental, cognitive, and social."[21] Moreover, and critical to establishing the motivation for architectural robotics, adaptive flexibility in the human species arose in response to "environmental instability."[22] As argued by Antón, Potts, and Aiello, the human species did not evolve in "a stable or progressively arid savanna" as suggested in earlier paradigms of evolution, but rather "in the face of a dynamic and fluctuating environment" composed of "diverse temporal and spatial scales."[23] So what distinguishes humans from other mammals is our adaptive flexibility, the capacity to "buffer and adjust to environmental dynamics."[24] The significance for architectural robotics is clear: the human species is superadaptive to "diverse spatial scales" and "environmental dynamics." This new paradigm for evolution, along with Lifton's concept of the protean self, suggest that we are prepared for, and can in all probability make use of, controlled physical reconfiguration of our built environments under those life circumstances that warrant their application.

From the diverse perspectives considered here, from Greek mythology, to psychology, to human evolution, we acknowledge the vibrant exchange between the dynamic world in which we live and the intimate and social nature of our being. Resisted or accepted, central to what it means to be human is to be fluid, resilient, and on the move. Borrowing Lifton's words, architectural robotics offers "many and new places" to individuals and groups of individuals facing wide-ranging challenges and opportunities. And specifically for architectural robotics, there are a number of ways to arrive at these "new places":

- You can select a place among programmed places to match your life need or opportunity.
- You can fine-tune and then "save" configurations to create "new places."
- You can allow the physical environment to anticipate your needs, and it will reconfigure itself a "new place" for you.

Architectural robotics provides its users the technological means for creating a careful balance between stability and flexibility in the given moment. Architectural robotics affords its inhabitants the capacity, borrowing Lifton's words again, to "modify the self to include connections virtually anywhere while clinging to a measure of coherence."[25]

In short, we and our physical environments are well matched: diverse, dynamic, adaptive and sometimes blurred. Architectural robotics aims to collapse further the boundaries that distinguish us from our physical surroundings when the conditions suggest the greatest benefit to the individuals and groups inhabiting them, within what is our increasingly cyber-physical, networked society.

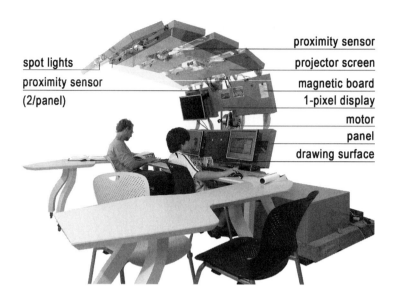

4.1
The Animated Work Environment (AWE), with its components identified.

4.2
Our developing, fully functioning prototype of AWE, showing its (unsheathed) aluminum-framed panels lifting "intelligently" as a user suddenly stands and departs.

4 IN AWE OF THE NEW WORKFLOW

In his review of *The Protean Self* for the *New York Times*, Richard Shweder captures the message of Lifton's book this way: that it's "better to be fluid, resilient, and on the move than to be firm, fixed, self-assured and settled."[1] While we do not envision (nor hope for) the entire built environment becoming architectural robotic, there are selective aspects of the built environment supporting particular human activities in which the protean pattern of reconfigurability may prove fitting. And perhaps more than any other physical environment, the twenty-first-century office workplace, with its demands for a "fluid, resilient, and on the move" workforce, warrants an architectural robotics response. The nineteenth century witnessed the evolution of the ample factory office from its beginnings as a dim corner of the factory floor occupied by a single, wooden desk. The twentieth century emerged with automation, labor management, and the modern office modeled after the assembly line: rows upon rows of steel desks sitting on the floor plate of an office tower. This highly structured office arrangement matured into the relatively organic "office landscape" of Herman Miller, which morphed into the office of late: Microsoft "Office," the thawing of the cubicles, and the expansion of hot desks, collaborative spaces, video conferencing, telecommuting, and LinkedIn. In Nikil Saval's *Cubed, the Secret History of the Workplace*, the author probes the physical office space, the nature of workers working in the office, and the relationship between work and home.[2] Saval takes comfort in what he calls the recent "quiet revolution" in the workplace: the swapping of office cubicles for office couches, the comforts of home transported to the workspace.[3] Still, the question remains: What becomes of work when it is no longer a discernible place dedicated to working life?

A response from architectural robotics comes in the form of the Animated Work Environment (AWE; figure 4.1) that was introduced earlier in this book and which serves as the case study for this chapter. An alternative vision to mobile and desktop computers, AWE is an interactive and user-programmable workplace environment that literally shapes the working life of multiple users colocated in a single, physical space, working separately and collaborating together. AWE supports and augments a twenty-first-century workflow, where knowledge workers are engaged in complex tasks requiring nontrivial combinations of digital and physical artifacts, materials and tools, and peer-to-peer collaboration.

While more of us are caught up in cyberspace, we nevertheless continue to find utility and value in working with and generating physical things.[4] We also maintain the need and desire for close collaboration with others, engaging together in complex work and leisure activities colocated in a single physical space. Indeed, ethnographic studies have consistently shown that people performing complex, creative tasks vigorously resist the "paperless office," preferring paper over computer tools for its ease in annotating, reconfiguring, and organizing information spatially and in shifting between storage, active use, and imminent use.[5]

In designing AWE, we sought to respond to these concerns by designing a workstation to meet two key goals: (1) *mixed-media use*, allowing users to use a range of digital and analog displays such as monitors, paper, whiteboards, and corkboards; and (2) *user-programmability* (reconfigurability), allowing users flexibility to rearrange digital and analog display areas to meet changing task demands.[6]

AWE is viewed as part of a growing tendency within computing research that is concerned with various cross-cutting issues related to working life, including use of multiple displays, managing mixed media, viewing health-care information, and, more broadly, practices defined as computer-supported collaborative work (CSCW).[7] In particular, AWE builds on contemporary workspace design and on prior developments of interactive workplaces with embedded digital technology, such as the Interactive Workspaces Project and Roomware.[8] Precedents from workspace design, however compelling, focus not on automated or physically reprogrammable spaces but mostly on beautifully designed furniture (without embedded electronics) that support conventional ways of working.[9] The informative precedents from human-computer interaction (HCI) and interaction design, meanwhile, are mostly defined by collections of computer displays, smart boards, and novel peripherals that create electronic meeting rooms. AWE sits between these two tendencies, at the interface between computer technology, architectural design, and automation where the physical environment (including display surfaces for paper) is also subject to physical manipulation.

An exemplar of the reconfigurable environment typology of architectural robotics that "makes room for many rooms," AWE affords wide-ranging workplace activities—individual work, coauthored work, prototyping, conferencing, presentations, serious gaming—to occur in the physical space of the common office cubicle or the smallest private office. AWE has two key physical elements: a user-programmable, robotic *display wall* equipped with an array of embedded sensors and digital peripherals, and a programmable *work surface* (composed of three table-like components), which is itself reconfigurable.

AWE's display wall is composed of six aluminum-framed panels (figure 4.2), each 5 feet wide, stacked on top of one another and connected as a column of panels by hinges (the original design called for eight panels). The hinges linking the panels are located close to the two extremes of every panel, which allows the linked panels to move much like a common metal watchband, but at the scale of a room. Each of AWE's panels is composed of an aluminum frame sheathed in plastic. The six hinged panels are actuated by electric motors. The motors are geared down via harmonic drives to enable the high-torque loads of the worst-case scenarios this system needs to handle, while minimizing the size of the motors. This makes for an elegant aesthetic design that aims not to overwhelm users with a display of massive motors.

AWE's panels serve as the structure for attached computer displays, lighting, audio, and other peripherals (see figure 1.2). The plastic sheathing of the panels provides a writable whiteboard-like surface, transforming the display wall into a large, configurable easel for writing on with common dry-erase pens. Beneath the sheathing of the panels are steel bands that allow for the display—interspersed with computer displays—of paper documents affixed by magnets, affording the use and generation of blended, digital, and physical media. The center segment of each panel houses LED lighting that can be switched on and off by users. This lighting can provide ambient lighting or serve as a super-low-resolution display of six large pixels, one pixel per panel, stacked vertically. This super-low-resolution display delivers background information on such things as the weather, the stock market, and such ambient information as the presence or activity of people in other affiliated workspaces. As informed by our task analyses elaborated in the next section, AWE was designed for up to three users working in close proximity, individually or collaboratively: we therefore equipped AWE with three computer displays so that each of three users could be engaged in separate computer activities, each requiring one display. As well, we designed AWE for an optional fourth display, smaller and lighter in size, mounted on the reverse face of the panel farthest from the base, so that one of the three users or potentially a fourth user could be working on the opposite face of the display wall when the wall is curled backward and toward the floor (as shown in figure 1.3, configuration 3, the lower-left panel of the figure). All computer displays are manually user-adjustable (figure 4.3) and are mounted following established ergonomic specifications. Notably, the research team first envisioned making the configuration of the computer displays intelligent via motors and sensors, but the team came to realize, through observation and simple reasoning, that sometimes some aspects of architectural robotics are best left to manual configuration, as in this case.

The three mobile, interlocking components of the work surface of AWE are, effectively, three tables of our design, set on casters that can be manually reconfigured ("programmed") by pulling them away from one another as three separate entities, nestling them together to make a smaller worktable composed of two tables, or a larger meeting table composed of all three tables, or alternatively forming the three tables into a U-shaped workspace for intensive, collaborative, cyber-physical creation (figure 4.4). Embedded with infrared (IR) proximity sensors

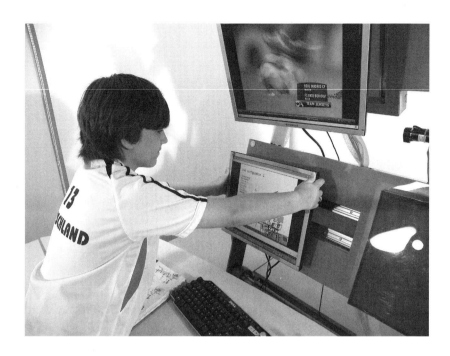

4.3
A younger user manually reconfiguring AWE's computer displays.

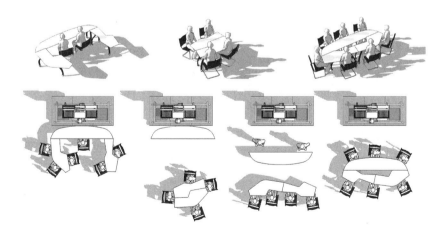

4.4
AWE's three work surfaces can be joined in different ways to match the dynamic workflow of users.

at their chiseled edges, the three table elements composing the work surface recognize each other on contact, becoming large-scale *tangibles* (*manipulatives*, in computer-speak)—a hands-on means of programming the physical configurations of the display wall by fitting the tables together as described. The manual and electronic means render AWE's work surface reconfigurable and reconfiguring.

While AWE is equipped in the way described here, we envision it as a platform equipped with still other digital and analog tools, varying and interchangeable, some of which have yet to be conceived by us or by others. In this way, AWE can be viewed as an open chassis: an open-source, cyber-physical system. And while we envision AWE as an open chassis, it can also be viewed as a manifestation of Christopher Alexander's concept of "compressed patterns" applied to the workplace rather than the home; that is, many workspaces, past and future, high-tech and low-tech, emerging from within one cyber-physical artifact.

CONFIGURATIONS SUPPORTING THE TWENTY-FIRST-CENTURY WORKFLOW

In a few words, AWE reconfigures to the demands of a twenty-first-century workflow that is dynamic and flowing. And "dynamic and flowing" defines not only working life today but also the traits of architecture that ensure its relevance tomorrow.

In this context, to imagine an architectural work of a particular type or functionality today requires (partly, at least) that it be something else, that it be *dissimilar*, not in the formal sense (of exhibiting a different shape) but *in the programmatic sense* (of accommodating "other places"). Consequently, we come to comprehend architectural robotics by embracing the memorable words of the iconic children's show, *Sesame Street*: "One of these things is not like the others." With respect to design specifically, philosopher Giovanna Borradori articulated it this way: "de-sign is always implicitly a project of historical-cultural disarticulation of the sign."[10] For instance, in coming to know what a home *is*, it's helpful to consider what a home *isn't*: it isn't (or, at least, shouldn't always be) a cell, a sacred space, a workplace, a gymnasium, or a school. But most of us know that the space of our home accommodates many of these "other places" in the course of a single hour, day, week, or month. Maybe the dynamic, flexible, flowing usage of physical environments lends to them the common word, *atmosphere*, or in Italian, *ambiènte*, which, with its prefix, *ambi-*, denotes that physical environments are as much defined by the mood they create than by the area or volume they envelope or the style of their design; that, within an interior environment, the inhabitants sense and, in fact, help generate a tension or play (*ambi-*) between themselves, the interior furnishings, their physical enclosure (of walls, floors, ceilings, etc.), and the outside world. While an *ambiènte* (a physical, interior environment) naturally contains fittings and furnishings in a space defined by the disposition of walls, it also includes all kinds of sensory experiences, memories, and events associated with the place and with all the people and things passing through it.

Inspired by this way of thinking, the Animated Work Environment assumes different physical configurations primarily by way of its embedded electronics and, particularly, its robotics. It is hardly the first workplace to assume different configurations; an ancestor to AWE existed in sufficient numbers in the eighteenth century, long before the ubiquity of electronics, let alone computers. The so-called Rudd's table (figure 4.5) is a wooden table of slender, exacting proportions with a work surface that seemingly levitates in mid-air. The table's two flanking drawers open not directly outward on fixed rails, but instead outward and also hinged away from the user seated at the table. Two relatively tall mirrors, the depth of the two hinging drawers, hinge upward from hiding. Offering at least two different workplaces within a single artifact, the Rudd's table is a remarkable feat of design, technology, and craft that my lab team and I aspired to in developing AWE.

4.5

A George III mahogany and marquetry Rudd's table, circa 1770 (reprinted with permission of Bridgeman Images).

While the Rudd's table offers two different configurations and their variants, limited partly by mechanical possibilities and the absence of electronics, the question for the AWE team and, more broadly, for architecture—embedded with robotics—is this: How many physical configurations ("other places") can users comprehend, and what physical form should each configuration assume? In broader terms, the question is: How do configurations lend themselves to our comprehension? A reply comes, for one, from Gestalt psychology, which teaches us that we structure our experience of the world by way of patterns. In perceiving external phenomenon, we complete what we judge is missing from them, and we associate them by proximity, continuity, similarity (or *dissimilarity*, to keep consistent with the previous consideration of *de-sign*), and still other "gestalt laws."[11]

This associative concept of Gestalt psychology is rooted in human evolution, and specifically in pattern recognition as an outcome of bipedalism. Standing on two feet, our pre-human ancestors obviously experienced a much wider world compared to when they were quadrupeds, living close to the ground; and the demands of making sense of this wider world, as has been argued, triggered the development of pattern recognition.[12] Given the centrality of pattern recognition for making sense of the environment within which we exist, and given its inseparability from human perception, memory, learning, and thinking, pattern recognition is one of the core problems of cognitive psychology and neuroscience, not to mention (coming full circle) the inspiration for Christopher Alexander's *A Pattern Language*, referenced here in chapter 2.[13] For computing, pattern recognition has become synonymous with machine learning (a topic we will return to in chapter 7), in which computers are programmed to discern patterns in data, whether these data are "data-mined" from a large database (as in *big data*) or sourced from the natural and physical environments and, sometimes also, the people and things inhabiting them (*ambient intelligence*). In this cursory consideration of pattern recognition, we again see a relationship between the animate and inanimate, people and machines, as well as between architecture (*A Pattern Language*), cognition, and computing.

My conception of AWE, and generally architectural robotics, finds validity in this breadth of research on pattern recognition. Early on in the development of AWE, my lab team and I dedicated ourselves to defining a finite number of patterns, or more precisely, cyber-physical configurations for AWE that we believed were, and later evaluated as, conducive to working life for an increasingly cyber-physical workflow.

Notably for AWE, and more broadly for architectural robotics, Leszek Rutkowski, a Polish researcher in computational intelligence, proved that the generally accepted model of pattern recognition holds up well under conditions in which the environment is not static but instead in motion (or what Rutkowski calls "nonstationary)."[14] My lab team and I envisioned AWE as specifically nonstationary: as an architectural environment capable of assuming a multitude of configurations supporting work activities of individuals and groups of individuals, colocated in one physical space, working with and generating digital and physical matter.

In developing AWE to support and augment the practices of workers in an increasingly digital society, my research team performed a number of qualitative, ethnographic studies led by team members with backgrounds in sociology and human factors psychology, focused specifically on survey and task analysis. We wanted to know what members of our target audience for AWE were doing when they worked and how AWE might support and augment these activities. And subsequently (as will be elaborated in this chapter), once we developed a functional AWE prototype, we wanted to observe how our target audience used AWE in representative work activities in order to further develop the prototype to support and augment these kinds of work.

A COMPELLING CASE OF INTERACTIVE FURNITURE AS A COMPARISON

Related to AWE is the compelling case of an interactive bench, *coMotion*, recently developed and evaluated by Erik Grönvall, Marianne Graves Peterson, and colleagues at Aarhus University (figure 4.6).[15] This upholstered bench, in its static state, affords users an opportunity to sit, lie, or lay down their belongings. Shape-shifting of the bench occurs by the rise and fall of four pairs of linear actuators, hidden from view under the upholstery, which act as the bench's legs. These actuators can raise or dip segments of or the entire bench, yielding different configurations designed to provide occupants a (not too) surprising jolt to make more space on the bench for others or alternatively to push sitters closer together or farther away from each other, among other possibilities.[16] To realize these different configurations, *coMotion*'s actuators are sent instructions by remote control and/or by input from how people are seated on the bench, as detected by six embedded force-sensors. In evaluating *coMotion*, Grönvall and Graves and colleagues applied six "sense-making" processes—*anticipating*, *connecting*, *interpreting*, *reflecting*, *appropriating*, and *recounting*, plus one of their own, *exploring*—as a structured means to observe different interactions between this robotics-embedded bench and its users.

Close cousins, *coMotion* and AWE are both larger-scale, cyber-physical artifacts with embedded robotics. Interactions with both artifacts occur through the manual selection of programmed configurations and through intelligent responses to user activity. Evaluations of both artifacts were performed in a structured manner that provided reportable outcomes. AWE and *coMotion* differ, however, in two notable respects. First, their purposes are different in tenor:

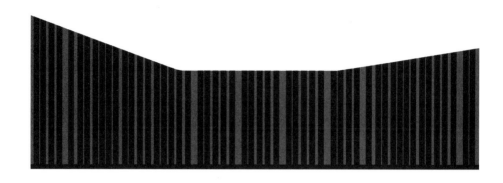
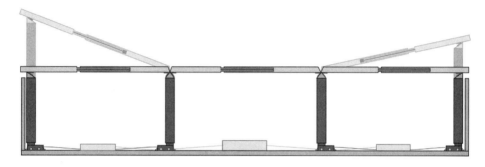
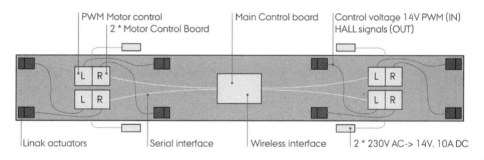

4.6

Technical drawings of *coMotion* (reprinted with permission of Marianne Graves Peterson, Aarhus University).

coMotion is meant to be playful in its physical reconfigurations, disrupting our expectations of a bench in a public place, while AWE's reconfigurability aims to support and augment workers' performance (productivity, creativity, collaboration, work satisfaction). Second, while the evaluation of *coMotion* by way of a finite number of sense-making processes provides an apt vehicle for making sense of the behavior of its users, AWE had to be evaluated for its efficacy in supporting the workflow of users. In these two respects, AWE is arguably a more complex, human-machine phenomenon. So while *coMotion* is compelling, careful, and whimsical in the best sense of architectural robotics, alone as a single piece of furniture it doesn't yet form an architectural robotic environment: this would require that it be part of a suite of such intelligent furnishings that interact with each other and users, and that it remember and even learn from its encounters. These traits render additional complexity to an architectural robotic artifact like AWE, which itself constitutes a space-making environment; or, the assistive robotic table (ART; presented in chapter 7), which constitutes the key piece of interactive furniture in an envisioned suite of intelligent furniture called *home+*. Fittingly, as elaborated in the sections that follow, the iterative development of AWE involved, early on, the surveying of knowledge workers and the observation of knowledge work practices, as well as, later on, the iterative evaluation of the developing AWE prototype through observations of users performing representative work tasks.

ETHNOGRAPHY AND AWE: LEARNING ABOUT THE NEW WORKFLOW

Early in the AWE research cycle, the AWE research team performed four hundred phone surveys with individuals in two relatively affluent and technologically savvy communities (Cambridge, Massachusetts, and Santa Monica, California) to better understand work practices among the expected target audience for AWE. The outcomes of the surveys confirmed our initial assumptions about AWE:

- Many work activities involving computing are not performed in conventional, desk-and-chair work environments.
- Computers and analog/physical tools are frequently used while performing the same task.
- Despite the use of both computers and/or analog/physical tools by the tech-savvy respondents, respondents sometimes prefer using physical tools rather than computers.

We conducted task analyses to examine user needs and preferences and to help generate design requirements for AWE. The task analyses involved 1.5-hour interviews with workers in their everyday work settings. The participants interviewed were workers who not only gather and process large amounts of information but also generate new "information products" (e.g., a webinar, an online workbook, an e-book) as outcomes of their work. To assess the use of information in various modalities—visual, spatial, textual, and numerical—participants consisted of four architects, four teachers, and four accountants. The interview data were analyzed with the goal of understanding how these workers gather, organize, store, communicate, and compose information—electronic and paper-based—using their current workplace technologies. We found the following:

- Most of the workers in our study used both paper and electronic information displays at every step of their work process.
- Electronic information processing technologies were frequently used along with paper.
- In often occurring collaborative work, earlier, informal work products were communicated to other workers electronically, whereas later, more formal work products were communicated using paper.

Overall, the survey and task analysis underlined the need for reconfigurable work environments like AWE, suited to a wide range of tasks involving simultaneous use of mixed-mode, physical and digital materials. Multiple users, computer displays, and computers were identified as important for AWE and their desired configurations suggested. The ability to support privacy in some situations, while allowing collaboration in other situations, was also identified as important to AWE.

Early in the development of AWE, the research team quickly built a full-scale prototype (figure 4.7). This activity was motivated by the working credo of graphic designer Bruce Mau when confronted with a wicked design problem: "Make mistakes faster."[17] Inevitably, human beings make mistakes, especially when confronted with unwieldy, underconstrained challenges such as envisioning the workspace of the future. Consequently, my lab team and I thought we would benefit from designing and building something of the AWE before we arrived at a sufficient understanding of its motivations, theory, and requirements. In this way, we could make inevitable design mistakes early on (*faster*) and learn from our mistakes as we came to design subsequent prototypes over a number of iterative cycles of (re-)design and (re-)evaluation. In less than two weeks, then, the team sketched out, at full scale, the AWE as we envisioned it at that early (premature, really) instant. As expected, most of this initial AWE prototype did not function by way of computer control and robotic actuation; instead, the lab team became *puppeteers*, controlling AWE's components by way of nylon threads and other kinds of manual manipulation. The video of a choreographed work activity involving workers (played by our graduate students) interacting with the early prototype in a manner convincing enough to be perceived as fully functioning, as if its different behaviors were actually being made controllable, interactive, and/or intelligent by way of computing. The method of this "magic" has a name in HCI research: *Wizard of Oz* (or WoZ, for short), in which the development of an artifact is guided by the investigators simulating (with low-tech means) an *apparently* fully functioning system in anticipation of its development as a properly full-functioning system.[18] From this early AWE, the team gained a sense of the scale and possibilities for the system.

The subsequent full-functioning AWE prototype embedded with robotics (figures 4.1–4.3, 4.8, 4.10), was strongly informed by the early WoZ prototype, the outcomes of our ethnographic studies, and by our consideration of ergonomic standards, workplace design precedents, and research literature and theory drawn from multiple disciplines. From the outset of the AWE research project, concurrent with the varied activities just described, the research team met weekly to develop two preliminary documents: the guidelines for establishing the number and physical specificity of AWE's configurations (and related aspects of the design); and design drawings, developed with CAD software, capturing alternative designs, each of these serving as visualizations of the developing AWE, informed by our guidelines for it. The guidelines included the following points:

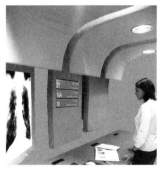

4.7
Our initial AWE prototype with mostly simulated "WoZ" functionality—a vehicle for making inevitable design-research mistakes, quickly.

- Multiple information displays are desirable and should be located in close proximity to one another.
- Displays of information that are digital, printed, or a mix of both should be located proximate to users.
- Space should be allocated for handwritten notations.
- Printed information used frequently by workers should be made very accessible—on top of horizontal and vertical surfaces close to users.
- Work environments should accommodate multiple people, sitting side by side or across from one another.
- Work environments should provide a large whiteboard and/or computer display for group activities.
- Work environments should provide a degree of privacy by blocking unwanted visual access and auditory intrusion.

In preparing the guidelines and associated design drawings, the team drew on the insights gained from all of the activities described here, as well as on the collected knowledge of the team members representing architecture, robotics, human factors psychology, and sociology.[19] Team meetings were characterized less by freewheeling and friendly brainstorming and more by critical and sometimes tense contributions and debate. Indeed, studies have shown that critically minded, frank contributions to group work are far more productive than uncritical brainstorming.[20]

THE SIX CONFIGURATIONS OF AWE AND THEIR USABILITY TESTING

A close consideration of the guidelines and the accumulation of design drawings informed the design of six standard physical configurations of AWE in support of individual and collaborative human activities that describe the workflow in a day across a continuum, from intensive work activities (e.g., *composing* and *presenting*) to less intensive work and leisure activities (e.g., *viewing* and *gaming*) that define the new work environment (figure 4.8). Each configuration is identified by a number, 1–6, as labeled in figure 1.3 and referenced in the text that follows. Notably, each configuration is not defined by a single activity, but rather by a cluster of activities envisioned for it. For instance, configuration 1 is envisioned to support *collaborating, composing,* and *viewing* activities, while configuration 4 is intended for *gaming, lounging,* and *playing*. To call up a particular configuration, users select one of six numbered buttons located just below the first frame from the base. Each configuration permits users to envision still more uses for it, and all six configurations can be gross or fine adjusted by IR sensors located near the ends of AWE's panels, as will be considered shortly.

4.8

Photographs of AWE in two of its six configurations: configuration 1 (top) and configuration 5 (bottom).

Configuration 1 (labeled "1" in figure 1.3, and shown in figure 4.8, top) affords intensive composing and viewing of electronic and printed information by one or two users. The focus in configuration 1 is on the three lowest screens, which can be positioned so that either (1) one or two users can focus on the same set of displays, with all three screens positioned closest to center, or (2) two users can work separately, side by side, with the two lower screens set apart.

Configuration 2 ("2" in figure 1.3) affords intensive computing by a single user who might elect to position the two lower screens toward the vertical center. When attached to the terminus of the reconfigurable wall, a privacy screen extends toward the floor to block visual access from behind the user. Physically, the privacy screen (which doubles as a projector screen for configuration 5) is a rectangular, fabric tarp with reinforced edges and magnets sewn in each of its corners. As needed, the magnetic projection screen is easily attached to projecting steel disks that define its corners. The privacy screen hangs freely from its two attachment points on the terminal panel of AWE. Also maintaining privacy is an envisioned, hinged leaf that can be folded from the work surface upward to provide partial visual access from the side, presuming that AWE is set with its other side near a wall. Should AWE be placed in a room where the wall is to the right of the user, the two outer work surfaces, both on casters, are easily repositioned to offer a like measure of privacy.

Configuration 3 ("3" in figure 1.3) affords composing by two individuals engaging in activities that do not require them to share the same intimate space. This might be the case where the two users are working alone on different pursuits or different aspects of the same pursuit and welcome the modest distance this spatial relationship creates between them.

Configuration 4 a ("4" in figure 1.3) affords two users working in the same intimate space, but back to back. This configuration suits two people gaming. It is also suited to working collaboratively, but unlike the side-by-side collaboration of configuration 1, this configuration better supports a scenario in which the collaborating individuals are working on different but related documents (say, pertaining to a single project) or are working on different aspects of a single document.

Configuration 5 ("5" in figure 1.3, and shown in figure 4.8, bottom) affords, most particularly, formal presentations requiring a projection screen. The projection screen is the same rectangular, fabric tarp described for configuration 2, attached in this case to four projecting steel disks to keep it taut independent of the position of the display wall. The work surfaces of AWE are repositioned and rotated 180 degrees to allow room for the presenter, a podium, and a pedestal supporting physical artifacts as part of the presentation. Lighting integral to AWE's panels is programmed to focus light onto the physical objects displayed or the speaker.

Configuration 6 ("6" in figure 1.3) affords leisurely viewing of videos or slide shows or group gaming presented on the projection screen. This configuration suits the playback of movies, satellite television, and other longer time-based media.

In developing these six configurations, the research team conducted usability tests in which users performed representative work tasks using AWE in one of its six configurations while analysts from our research team observed and recorded their behavior. The usability tests of AWE were novel for HCI: rather than testing shorter-term tasks, these usability tests for AWE tested for long-term tasks in which users were required to access large amounts of multimodal information and integrate this information into a complex product. Consequently, we had users perform tasks of relatively long duration (two hours). We selected two different complex tasks, architectural design and tax preparation, to gain insights from two different modes of work activity involving both printed and digital matter. For the design task, eight senior students from Clemson University's School of Architecture created preliminary designs of a multifamily residence. For the tax-preparation task, four participants with measurable tax-preparation experience completed a complicated tax return. During each test session, users were asked to verbalize their thoughts. Using videotaping and a real-time coding program, the research team recorded the focus of attention of each user working within the AWE workspace throughout the test session. Our team's analysis of the data focused on spatial and temporal components of how users used the paper and computer displays of AWE. To capture and communicate a key aspect of our observations, the research team prepared "heat maps," which capture users' relative frequencies of use of the various paper and computer display locations that AWE offers (figure 4.9).

Results from this research activity involving configuration 2 provide a sense of what can be learned from usability testing of architectural robotic artifacts such as AWE. Somewhat surprisingly, all twelve participants used paper regularly in both the design and tax-preparation tasks *despite task constraints encouraging computer-based work*. Many participants made extensive use of AWE's horizontal tables for spreading out paper spatially. People switched between paper use, computer use, or combined paper-computer use at different stages of their tasks, which supports one of the key findings of our

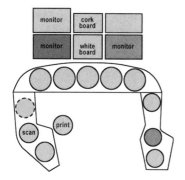

4.9

Heat map communicating frequency of use of AWE's paper and computer display locations for a "very-high paper user."

4.10

Fine-tuning AWE by gesturing toward IR sensors.

task analysis considered earlier. Our usability findings also began to validate that AWE achieved one of its primary design goals: to facilitate flexible use of paper and digital media. With regard to the goal of better integrating nondigital displays into knowledge work, AWE's vertical, nondigital display spaces proved popular among participants. Two thirds of the participants used these displays, mainly for short-term, "hot storage" information storage. Overall, our usability test findings regarding use of mixed media provide quantitative support for the qualitative findings of our task analysis and other ethnographic studies of knowledge workers.[21]

From a broader viewpoint, AWE promises to automate some higher-level work activities where, like the factory floors of today, fewer workers may be employed. As reported by RoboticsTrends.com in September 2014, a survey of Harvard Business School alumni found that 46 percent would rather have robots perform their company's labor than hire people to do the same work.[22] Robots are welcome, robots are coming, and AWE promises to be the kind of robot positioned to play human-centric and business-centric functions.

Users are not limited to the six programmed configurations presented here: configurations can be fine or gross adjusted. These adjustments are made by a user gesturing toward the sensors embedded in the four innermost panels of the fully-functioning display wall, one IR sensor on the left side and one IR sensor on the right side of each panel (figure 4.10). These sensors provide programmable functionality of the display wall in response to sensed, real-time proximity data. The sensors are configured in an attract/repel mode, whereby gesturing one's hand in proximity to the sensor located on the right side of the panel draws that panel toward the user, while gesturing one's hand in proximity to the sensor on the left side of the panel repels that panel. This interaction is designed to allow users to program the shape of the display wall according to their particular needs. New configurations of the system can be "saved" for future recall. Additionally, AWE exhibits that bit of intelligence prescribed by Negroponte (as considered in chapter 2): its display wall automatically responds to unanticipated and unexpected movements of its users, as when a user suddenly stands up from a seated position (see figure 4.2). In this instance, AWE's column of IR sensors acts reflexively, repelling the wall upward and away from the sudden approach of users. The column of IR sensors, operating in these ways, was found by our team to be particularly intuitive and effective for altering and fine-tuning functions.

A SCENE FROM THE MARRIAGE

Continually, my research team and I wove the threads of our many meetings together as a written scenario, a narrative vehicle common in HCI research that defines the desired interaction between the artifact and likely users—an aspirational story of their ideal union. In their brevity, scenarios tend to be more haiku poem than epic. They are meant to capture a few characteristic moments of interaction: scenes from a marriage, not an elaborate life story. Our scenario for AWE is this:

> Martha approaches AWE, currently in a resting state. She inserts her USB drive into the port on the front face of AWE, transitioning AWE into standby mode. The USB drive contains a saved AWE configuration from her previous use. AWE's software recognizes this bit of information and moves to the saved configuration (configuration 1, supporting primarily collaborating, composing, and viewing activities). Seated before AWE, Martha opens a partly finished presentation document file (an upcoming conference talk she will give next week) on one of AWE's computer displays and the corresponding conference paper on another. She takes her notes outlining her planned modifications to the talk from her bag and spreads them on AWE's horizontal surfaces. She begins to work on her presentation, using both the digital and hard-copy information.
>
> Martha realizes she is not being very productive and feels she would concentrate better in a closer, more intimate environment. In AWE's control window, she toggles configuration 2. Quickly and smoothly, AWE's surfaces fold to create a more confined space. Still not completely happy with the resulting configuration, with a single press of AWE's control window, Martha places AWE in fine-tuning mode and subsequently extends her hand toward the infrared sensor located at a hinge point between two panels to draw these panels closer to her. Martha keeps her hand extended until she has AWE configured in the shape she imagines. She continues work on her presentation, making steady, satisfying progress. Suddenly, however, Martha remembers that she may have forgotten to bring with her an important document left in her car outside the office building. She stands up quickly, grabbing her key to go outside to her car. AWE's proximity sensors detect Martha's unexpected activity and, responsively, AWE's panels move upward and away from Martha to allow her an unobstructed path to the door. After several minutes of inactivity, AWE transitions back into standby mode.

ROBOTICS SUPPORTING THE TWENTY-FIRST-CENTURY WORKFLOW

To achieve the kinds of behaviors described in the scenario and suggested by our wide-ranging research activities, we implemented the design goals of user programmability and reconfigurability by lending AWE a capability for robotic movement. To assume its different configurations, whether programmed by us, user specified, or intelligent, this architectural robotic workspace required a different kind of robotics: one that moves across physical configurations in a manner that is not alarming to users and, moreover, comprehensible and even graceful as perceived by them; one that supports wide-ranging work activities that are less rather than more repetitive, occurring in a space where the interaction of machines and workers is highly unstructured, unlike that of the highly choreographed factory floor.

For the layperson outside the field, robotics is not of the kind exhibited by AWE. Instead, robotics is commonly recognized as coming in two flavors: *industrial robots* that perform laborious tasks requiring repetition, precision, and (very often) strength (a typical one is shown in the far-left image of figure 4.11 in the next section), and *humanoid robots* that, because they look and behave a bit like us, capture our imaginations (as captured in popular film as well as science fiction literature, from which the name *robot* originates).[23]

But why not use industrial or humanoid robots to support working life in the way that AWE aims to support working life? Perhaps a robot arm from the factory floor might perform a necessary function in the administrative office of your department? Or maybe a humanoid robot can be commanded by the office staff to help it perform its duties? To understand why and how the robotics embedded in AWE are well suited to its intended applications, we need to know something more about its competition, about industrial and humanoid robots, by *dissimilarity—by what AWE is not*.

With respect to industrial robots, first, industrial robots perform well in structured environments such as the warehouse and the factory floor and not very well in unstructured environments where we conduct our everyday lives, such as our homes, schools, workplaces, and in hospitals where anything and everything can happen—*life as it is*. Industrial robots are programmed to perform singular functions repeatedly, and they perform these tasks by way of limited (stiff) movements of their typically rigid, massive components. Moreover, the unyielding movements of the large physical masses typical of industrial robots are sometimes the cause of mishaps, injuries, and (rarely, but reportedly) fatalities.[24]

With respect to humanoid robots, they arguably have a long way to go before they become the safe, reliable, and useful assistants promised us as we conduct our lives in the everyday places just mentioned. Humanoid robots, for one, have difficulty doing things we do instinctively or with little effort, such as picking up after themselves or (with disastrous results, you can imagine) picking themselves up after a fall. Named for robotics pioneer Hans Moravec, *Moravec's paradox* describes this ineptitude. According to Moravec's paradox, it is "comparatively easy to make computers exhibit adult-level performance on intelligence tests or playing checkers," but very "difficult or impossible to give them the skills of a 1-year-old when it comes to perception and mobility."[25] Undoubtedly, in recent years, there has been increased momentum in humanoid robotics research and overall incredible findings from this research domain. Nevertheless, much needs to be accomplished to meet safety requirements in unstructured environments and to meet (your Aunt Ida's) expectations for what such a robot servant can do. All signs suggest that humanoid robots will increasingly become integrated into some aspects of our everyday lives; and, for design researchers in robotics and allied fields, there is much to learn from our peers working in this research domain. Still, there are doubts about when or if an affordable, reliable humanoid assistant of vast motor control and intelligence is coming to anyone's home soon.[26] If and when it does, they are welcomed, together with architectural robotics, as citizens of envisioned *ecosystems of bits, bytes, and biology,* as will be elaborated in this book, particularly in its epilogue.

In less structured environments such as homes, schools, and medical facilities, the more intimate and nuanced interactions between people and robots will require of robotics two capacities that most robots operating today currently lack: expanded mobility and increased compliance. With respect to mobility, architectural robotic environments, especially given their relatively large scale, must move physical mass through space in a manner that is, for its users, safe, supportive, and graceful (or at least not distracting). And as physical mass is moving through space in architectural robotic environments, the robotic aspects of these environments must be compliant (resilient, "give," "yield") upon contact with another physical mass (such as our property or our bodies). Over millions of years of evolution, the human species and other living things have developed such instinctive and very precise capabilities, operating by way of a complex network of physiological systems. Inspired by nature, but not imitative of it, architectural robotics depends on advanced capacities for motion as well as heightened compliance in order for it to support and augment human activities.

KINEMATICS (TRAJECTORY/MOTION PLANNING)

Here we attend to the matter of choreographing the graceful movements of architectural robotics, waiting until chapter 7 to consider the matter of compliance. And whether graceful or not, the movement of robotics and other moving bodies is described by *kinematics*, rooted in the study of geometry and differential geometry. More precisely, differential *kinematics* describes how a robot moves from one physical configuration to another configuration to arrive at its *goal*—the geometric models for physical, spatial configuration. If I use my arm as an example, lifting a book from my desk requires that my wrist, elbow, and shoulder move cooperatively. Over the short time it takes me to achieve this simple task, the position of each of these joints, and the angles between them, define the plan of motion (*path* or *trajectory*) for my arm. If I replace my arm with an equivalent robotic arm, I can translate these positions and angles to the robot's motors to mimic the motion of my arm.

To move from one AWE configuration to another, as introduced earlier, a user selects one of AWE's six programmed configurations by selecting that configuration from a simple "push-button" interface; alternatively, if AWE is operating more intelligently, infrared sensors on the display wall infer information about what's going on both within and in the external environment surrounding the AWE. AWE's actuators (its motors) transfer its mass (the various panels composing its display wall spine) from one configuration to another. The use of sensors and actuators to reconfigure a robot according to a desired intention (or plan) is defined as *feedback control*.

Forward kinematics is one of two ways to define the trajectory of my arm: we start with the joint at the base of the limb (my shoulder) and move down the limb to the descending joints (from shoulder, to elbow, to wrist) and wind up at the hand, the "end effector" that I'm aiming to position to lift my book from the table. As the term implies, *inverse kinematics* has it the other way around: I start by determining where I want my hand (the end effector) to go to lift my book from the table, and then I figure out how to get my joints, over a short duration of time, to achieve that goal for the given *task*.[27]

To begin to consider kinematics for architectural robotics, we need to know something first about how the robotic arm you might find on a factory floor characteristically moves (its *behavior*). Consider the typical industrial robot you might find on an assemblyline floor for car production. It's something like our arm, composed of a small number of serially connected segments, or *rigid links* (typically four to six)—a mechanical assemblage well matched to "pick-and-place" applications such as installing a windshield into the steel frame of an automobile in production. We describe such a robot as *vertebrate*, as it is composed of segments connected by joints much as our spinal column is composed of twenty-six bones connected by cartilage (refer again to figure 4.11, the far-left image). Such robots are controlled at each joint, which commonly acts as a hinge, much as our elbows do.

For each of its joints, the vertebrate robot exhibits 1 *degree of freedom* (DOF), meaning that the configuration of that joint is specified by a single variable. (This is akin to our elbow, which, operating normally, bends in one direction, as if there was an axle inserted through it, serving as its fulcrum.) As commonly defined, a *robot* is a physically reconfigurable system that has multiple degrees of freedom across it and is (at least partially) computer controlled.[28] In the case of the industrial robot being considered here (as in our elbow), the limit of 1 degree of freedom lends it both rigidity and strength. For most tasks, the kinematics for

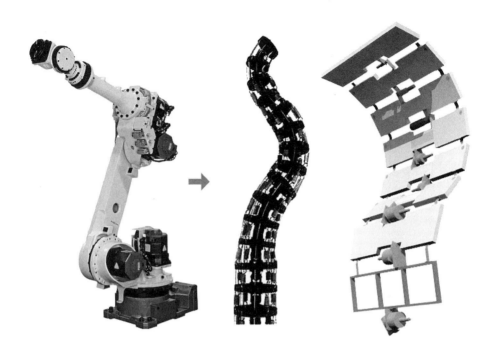

4.11
A vertebrate robot that might be found in industry, and two essentially invertebrate robots from the lab: the elephant's trunk and AWE.

this kind of robot is not overly complicated; there are not so many permutations of possible behaviors for such a robot, given its limited number of rigid links and the limitation of 1 degree of freedom at each of its joints.

But kinematics gets altogether more complicated (and more interesting) when the numbers of joints in a rigid-link robot increases beyond what is necessary for the task. Robots having somewhat more than 6 degrees of freedom—that is, more degrees of freedom than the industrial robot just considered—are called *redundant*, implying that there is more than a single pathway for the robot to assume a given (end effector, or hand) task location. In the more technical terms of robotics researchers, for a redundant robot, the degrees of freedom in the *configuration space* greatly exceed the degrees of freedom of the *task space*. In sum, a *rigid-link* robot has one way only to arrive at its goal configuration, while a *redundant* robot has more than one pathway to arrive at its goal configuration.

The display wall of AWE was designed initially to have eight panels, connected by eight joints, each having 1 degree of freedom, or 8 degrees of freedom collectively. Exhibiting 8 degrees of freedom, the display wall is classified as kinematically redundant for all the tasks we've designed for it, as all of these tasks require 7 or fewer degrees of freedom. Our final prototype has six panels; nevertheless, as built, this display wall is still classified as kinematically redundant for all the tasks we've designed for it, as all of these tasks require 5 or fewer degrees of freedom. In the simplest of terms, AWE's display wall, both in our initial design and as prototyped, has more degrees of freedom than it requires to move between any of the configurations we've designed for it: it has many ways "to go from here to there." In this way, the skeletal system of the display wall and its kinematics is like an *essentially invertebrate* snake having a larger number of links that bend at distinct and well-defined points (figure 4.11). As a kinematically redundant robotic surface, the display wall of AWE is a novel robotic surface—unique, to the best of our knowledge.

An unusual aspect of the display wall kinematics problem is that the regulation of the position and orientation of the tip of the wall (the end effector) is rarely the primary consideration, as is the case in the research literature concerning kinematically redundant robots (and as in the case of our elementary kinematics example involving a robot arm, introduced earlier).[29] For the display wall, the position and orientation of the three computer displays on the two lowest panels represents the primary task. Why is this so? Because for most configurations of the display wall, the computer displays need to be positioned so that AWE's users can work effectively with them (view them in the best way). Consequently, the two panel surfaces closest to the floor that frame and support the three computer displays are the primary consideration in this kinematics resolution problem, not the orientation of the display wall at the tip of its most extreme panel. This tip of the display wall only becomes important to the kinematics resolution problem, for instance, when (or if) spot lighting, mounted at the tip of the wall, needs to illuminate an object of considerable interest or otherwise track a speaker pacing in the room during a public presentation using AWE.

After the team converged on six configurations for the display wall, the next design challenge was to determine how the display wall will move between the six configurations in the safest, most graceful manner. In planning these various paths between the six reference configurations (trajectory planning), we adopted an approach in which the display wall moves with respect to a desirable, guiding pathway while still satisfying the primary task constraints. With this approach, the display wall moves between its six different configurations along pathways we judged to be safe, elegant, and manageable, given the torque of the specified motors we used at each joint between panels. We explored different guiding pathways for different configuration changes, as one guiding pathway alone could not serve as the guide for the safe, elegant, and efficient choreography of the display wall across the various configurations unfolding in different sequences, depending on how users are working with AWE. Eight different guiding modes were ultimately selected, inspired mostly by the forms and behaviors of living things—the cobra, the sequoia, the ostrich, the elephant's trunk, and other familiar things drawn from nature—to enhance the perception of graceful, natural movements in the display wall.

Let's observe one example of what this looks like, moving between two configurations of the display wall. Figure 4.12 captures the results of this approach as a comparison of two alternative pathways for moving from configuration 2 (supporting a user mostly composing) to configuration 5 (supporting someone primarily presenting). Note that in the figure, the display wall is viewed from its side (i.e., in section). Each image (left and right in the figure) represents the trajectory of the display wall in a series of timed sequences (like the well-known motion-capture photos of Muybridge). The sequence begins at configuration 2 on the lower left of the figure and moves to configuration 5 on the upper right. In each sequence, note as well that the section-line grows lighter in saturation (intensity) as time passes. Comparing the two sequences as shown in the figure, recognize that they differ in guiding configuration, so that their pathways of movement between configurations 2 and 5 are correspondingly different. The more favorable condition, the one we ultimately used for moving between configurations 2 and 5, is the one shown in the image on the left—the path guided by the elephant's trunk—as it provides an elegant pathway that also moved the

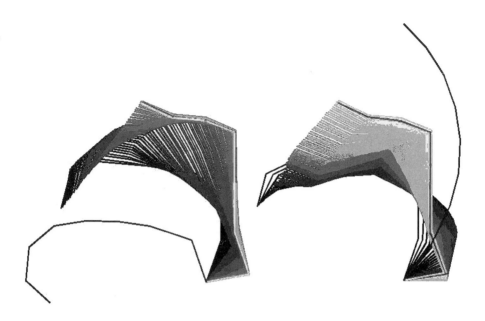

4.12
Trajectory planning from configuration 2 to configuration 5 using two different guiding modes, represented as a series of singular bent lines denoting the wall's profile over the course of its movement.

top and middle joints of the display wall into position relatively rapidly, reducing the torque on the bottom motors.

BEYOND THE SCREEN AND INTO THE CYBER-PHYSICAL WORLD

The network of increasingly powerful and inexpensive personal computing devices is revolutionizing many aspects of human existence, connecting individuals worldwide and making accessible to them vast amounts of information and opportunities. At the same time, our increasingly digital society is alarmingly characterized, at the workplace, home, or school, by a solitary user facing a computer screen of whatever size, accessing digital information within a static, physical environment. While people are increasingly caught up in cyberspace, they nevertheless continue to find utility and value in physical artifacts, materials, and tools, not to mention other colocated people.

An exemplar of the reconfigurable environment typology of architectural robotics that "makes room for many rooms," AWE is composed of novel robotic, digital, and analog components, meticulously designed to enhance purposeful human activities. Specifically, my lab team and I strove to realize in AWE a cyber-physical environment sufficiently flexible to support and augment contemporary, complex, and dynamic work practices—some of these known, some of these unexpected, and still others yet to be discovered. In AWE, users are not in the office, not at home, but in something that is *in between and other than*.[30] We envision this *in between and other than* environment as including AWE, smaller interactive and intelligent peripherals it grounds and connects, digital and printed tools and the products it helps generate, human and robotic workers of all kinds, the built structures that shelter them, and connectivity to similarly outfitted ecosystems and to the Internet, when and where needed.

5   THE RECONFIGURABLE ENVIRONMENT

Essentially Alexander's concept of "compressed patterns," where all the functions of a typical house occur in the space of a single room, the reconfigurable environment (figure 5.1) is a malleable, adaptive environment specifically dependent on moving physical mass to arrive at its shape-shifting, functional states supporting commonplace human activities. The reconfigurable environment is characterized by a continuous, compliant surface that renders a room relatively soft compared to the conventional rectangular room. In a reconfigurable environment, (most) everything across its continuous surface is capable of physically reconfiguring to create various configurations that evoke yet transcend walls, ceiling, floor, and furnishings, accommodating the activities of its inhabitants.

As with all three typologies of architectural robotics defined in this book, the reconfigurable environment is subject to "shared control" between the user and itself: the environment is not strictly intelligent, nor do users commandeer it. There is, instead, a degree of control, along a sliding scale, so that users are best served by this kind of human-machine rapport. This means that while the reconfigurable environment knows something about its component parts, it knows only a little something about its inhabitants and reconfigures itself fittingly.

The Animated Work Environment (AWE), particularly in its display wall, serves as an early design exemplar of the reconfigurable environment that literally gives form to the working life of multiple people, collaborating together in a single physical space. Moreover, figure 5.1 is meant to capture the essential character of the reconfigurable environment typology. Each of the three cells in the figure represent a brief period in the everyday life of the inhabitants living within this single volume, as its continuous surface morphs to allow them to play, lounge, eat, and speak. *The remarkable feature of the reconfigurable environment typology is that a single physical space makes room for multiple functional places supporting and extending human activity: "one space, many rooms."*

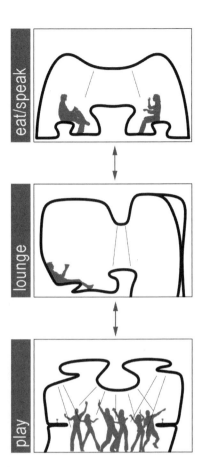

5.1
Diagram of the reconfigurable environment typology (diagram by the author).

The Reconfigurable Environment

II DISTRIBUTED

6 LIVING ROOMS

To live in the world as in an immense museum of strange things, of curious variegated toys that change their *appearance*....
—Giorgio de Chirico, quoted in "De Chirico in Paris, 1911–1915"

Maybe you can remember an instant in your life when some interior environment familiar to you (your bedroom?) or at least welcoming to you (a vacation retreat?) was momentarily transformed into the unexpected—a place *alive* to unforeseen prospects. Perchance you were a guest at that old inn by the sea. Did you notice inside its parlor, one night, under hazy light, a shift in the position of the *bureau plat* or a change in the decoration carved into the table's legs? Or maybe you frequently commute down a dimly lit road in the country or the city: did that vacant farm house or row house ever seem ghostly to you? Or can you recall, on a late summer night, arriving at your own home, and something about it unsettled you: the shadows and a defined patterning of light moving across the walls and ceiling, formed by the headlights of a passing car? If none of these, then surely you can remember an occasion from your childhood when the familiar contents of your bedroom changed their appearance, as they did to the delight of the painter Giorgio de Chirico?

If you answered "no" to any or to all of these, you should not be surprised that more than a few painters, authors, psychologists, and filmmakers have answered in the affirmative. In literature, Edgar Allan Poe comes quickly to mind, particularly his House of Usher with its "black walls" and "vacant eye-like windows."[1] As reported by the narrator of this famous Poe story, upon entering its shadowy rooms, "I wondered how ... the objects around me," familiar "from my infancy," were capable of "stirring up... fantasies."[2] Curiously, many times it is the places most familiar to us or at least the places intended to be most welcoming to us that are perceived as sinister or, in a more cheerful light, fantastical (figure 6.1).

We won't dwell on the darker side of this strange, perceptual puzzle: the "unhomely" of Poe, Freud, and E. T. A. Hoffman, adroitly considered for architecture by Anthony Vidler.[3] Quite the opposite for architectural robotics, we seek *homeliness in the fantasy*, an ensemble of furnishings that, independently and together, somewhat miraculously (with the benefit of creative design and computational means) transform themselves, supporting and augmenting our everyday lives. *Living rooms*, living with us.

Partly because it is a time-based medium, cinema is especially capable of capturing the brighter, wondrous side of the living room. In Jean Cocteau's film *La Belle et la Bête* (1946), the obscure vestibule walls of the Beast's castle are lined with living, gesturing candelabra: torches held by human arms that illuminate

6.1

A hauntingly beautiful domestic space for the lady of the nineteenth century. *Boudoir in the Mansion House, City of London*, ca. 1897 (photograph by Alfred Newton & Sons, reprinted with permission of HIP / Art Resource, New York).

the pathway for those who enter the castle (figure 6.2). In Peter Greenway's *The Draftsman's Contract* (1982), stone statues assume different static poses, and interior furnishings change form and assume different locations in the course of the draftsman's meticulous efforts to prepare drawings of the estate that contains them. More recent, and far lighter in tone are Pixar's *Toy Story* films, in which toys come to life and lead adventures, but only whenever human beings are absent. (Maybe Pixar has, among its staff, a few admirers of de Chirico.)

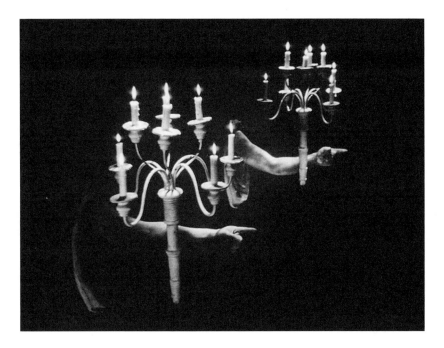

6.2
Living, gesturing candelabra. Film still from Jean Cocteau's *La Belle et la Bête*, 1946 (source: http://johnguycollick.com/la-belle-et-la-bete-1946/).

Among painters capturing such domestic eccentricities, de Chirico is joined by many of the younger Surrealists in Paris that he inspired and mentored. The most palpable example is Salvador Dalí, who, in works like *Weaning of Furniture Nutrition* (1934), conflated his memories of furnishings and a close, human relation into a near-photographic oil painting. Of painters commingling the animate and inanimate, Dalí may be, however, too obvious an example or at least one that chose to volunteer few and nebulous explanations about his artistic motives and processes. In this respect, de Chirico was the more generous, offering several essays and a novel. This novel, *Hebdomeros*, traces the movements of its fictional protagonist, of the title's name, through rather ordinary, urban streets, public spaces, and buildings. In this mundane, everyday environment, an objective understanding of the place mixes with a "metaphysical" one, of the memories and vague associations of Hebdomeros.[4] In de Chirico's essay, "Statues, Furniture and Generals," the artist articulates this mixed perception of the built environment as a physical existence and as something beyond ("meta") the physical:

> Perhaps you have occasionally noticed the singular aspect that beds, mirrored armoires, armchairs, divans, or tables have when you see them suddenly in the street, amidst a décor in which we are not accustomed to seeing them, such as happens during a move, or ... before the doors of shopkeepers and second-hand dealers who display the principal pieces of their merchandise on the sidewalk. Then the furniture appears to us in a new light; the pieces are dressed with a strange solitude; a great intimacy is born between them, and one could say that a strange happiness floats in this narrow space that they occupy on the sidewalk, in the midst of the ardent life of the city and the hasty comings-and-goings of men....[5]

As experienced by Hebdomeros, or as painted by his inventor, de Chirico (figure 6.3), a roomful of furniture is far more than a physical fact. These inanimate things, organized as a room, so familiar and central to our everyday lives, have a life of their own, at least in the way we perceive them. And how we perceive the world, remember, is not as directly connected to the world itself as we insist. With respect to the visual perception of our surroundings, we easily forget that "the brain must invert the image perceived, create a single perception out of a doubly experienced sensation, construct the third dimension, and then adds distance to complete the space."[6] Hardly a direct pathway between the physical world and our visual perception of it. Moreover, our visual perception of our surroundings is structured by our prior experiences and judgments: *our way* of seeing the world. As Immanuel Kant maintained, our perception of the world "must conform to our a priori knowledge" of it—an assertion since validated by cognitive science.[7] Pondering the implications of perception to specifically the experience of architectural works, Mallgrave elaborates that architecture "is a more deeply embodied phenomenon than the merely visual; it deals with many more sensory and subliminal dimensions (spatial, materials, and emotional) and therefore engages many other areas of the brain."[8]

In fewer words, architecture is omnisensorial, and our perception of it is complicated and complex.[9] Traversing the wide-ranging intellectual landscape of "lively" buildings, rooms, and furnishings early on in this chapter is meant to show that there is more to the built environment we live in than the static, fixed reality of it. This brief itinerary opens the doors, so to speak, to the concept of architectural robotics as a reconfigurable space, as a *living room* that has occupied the thoughts of a great many of us for a very long time.

A EQUALS B

The dream house must possess every virtue. However spacious, it must also be cottage, a dove-cote, a nest, a chrysalis.

—Gaston Bachelard, *The Poetics of Space*

How wondrous is the human capacity to perceive and misperceive, to imagine this world and envision another, all within the space of a room familiar to us. How wondrous, too, is the human curiosity to examine and sometimes reveal the wonders of the natural world, and in one early, documented instance, to observe the logic in the seemingly illogical, *that one thing can be another thing*, as in architectural robotics. In fundamental terms, this chapter is a response to a fundamental question: How can things belonging to a larger system, something as familiar as insects or furnishings, reconfigure themselves individually and collectively to be something else—a question that perplexed Aristotle when observing the butterfly's life cycle, which then led him to thinking about, of all things, the built environment.

In the *Generation of Animals*, Aristotle expressed his struggle to "put together" the workings of metamorphosis: specifically, how a distinct entity like a caterpillar could transform itself into something else, a butterfly.[10] After some false attempts at understanding this strange process, Aristotle curiously equated metamorphosis, the form-making found in nature, with the form-making that results from human labor: "As far as the things formed by nature or by human art are concerned, the formation of that which is 'potentiality' is brought about by that which is in 'actuality'; so that the Form [sic], or conformation, of *B* would have to be contained in *A*."[11] According to Aristotle, the process of metamorphosis demonstrated that neither *A* (the caterpillar) nor *B* (the butterfly) are

6.3
Released from the confines of the home, furniture mingling in the wild. Giorgio de Chirico, *Furniture in the Valley*, 1927 (reprinted with permission, © 2014 Artists Rights Society [ARS], New York / SIAE, Rome).

sufficient, "whole" entities in and of themselves; they are, instead, mutually dependent—bound, by definition—with A anticipating the arrival of B, and B following from A.[12] Later in the same work, Aristotle was more explicit about the process of metamorphosis, first equating it with the workings of the mannequin, and second with the building of a house:

> It is possible that A should move B, and B move C, and that the process should be like that of the "miraculous" automatic puppets: the parts of these automatons, even while at rest, have in them, somehow or other, a *potentiality*; and when some external agency sets the first part in movement, then immediately the adjacent part comes to be *in actuality*. The cases then are parallel: ... it is the external agency which is causing the thing's movement ...; in another way, it is the movement resident within which causes it to move, just as the activity of building causes the house to get built.[13]

Aristotle here makes a peculiar analogy between the natural process of metamorphosis, the mechanics of automatons, and building construction. At a much larger, physical scale, and much more contemporaneously, Aristotle's "building" analogy finds evidence in the Italian "morphological model" of the 1960s, by which the urban forms of cities evolve according to a process of formal transformations accumulating over time. Of course, architectural robotics doesn't operate at the evolutionary time of urban metamorphosis (measured in centuries) or the evolutionary time of nature (often measured in millions of years), but in real time, the real time of everyday human activities as they unfold in homes, schools, hospitals, workplaces, and other familiar places. But by analogizing "building the house" with the metamorphic process, Aristotle has provided us with a working concept for how architectural works might reconfigure. Much like the stages of metamorphosis described by Aristotle, architectural robotics is conceptualized, first, as a sequence of physical configurations, rather than a static entity at rest; and second, its transformation (its metamorphosis) results from "some external agency," such as the detection (sensing) of a change in the needs or circumstances of its users.

As for Aristotle's automaton analogy, the automaton is the pre-robot: the robot before electricity. In fact, one third of Lisa Nocks's history of robotics, *The Robot: The Life Story of a Technology*,[14] is devoted to automata (mostly those resembling human beings and animals), presented in her book as the precursors of their electronic offspring (for Nocks, mostly humanoid robots). Indeed, automata have had a long history, and their life-like movements fascinate us still. As English professor Prof. Kenneth Gross argues in *The Dream of the Moving Statue*, "the idea of motion in ... [the] inanimate ... is a concept of a sort so basic that we can hardly call it a metaphor."[15] In Greek, the word *automaton* most literally means "acting of one's own will" and makes its first appearance in Homer's *Iliad* to describe the autonomous movement of doors and wheeled constructions. Not until the 1940s, however, with the rise of General Motors, does *automaton* become the more practical *automation* used to describe the labor-saving control systems for the operation of industrial equipment. Architectural robotics makes something productive of all of this confusion: it falls somewhere in the middle of the continuum bridging the automated or mechatronic and the intelligent machine, drawing advantages from both.

In his *Ten Books of Architecture* (ca. 15 B.C.), the first significant consideration of architecture, Vitruvius echoed Aristotle's association of nature, automatons, and architectural works. In fact, Vitruvius established the design of automatons and machines (broadly defined) as *integral* to the architect's practice, devoting book X explicitly to their design and construction (after devoting the preceding nine books to establishing an idealized, static conception of architecture). As did Aristotle, Vitruvius also likened the terms *organic* (*organa*) and *mechanic* (*machinea*) such that "*machina* and *mechanicus* (from the Greek *mechos* - a means, an expedient, or a remedy) referred to any kind of contrivance; *organon* came from an archaic term, *ergon*, work. It follows that [for Vitruvius] the Latin *organicus* did not mean anything different from *mechanicus*, something done by means of instruments, indirectly."[16] This early and persisting identification of *organic* and *mechanic* implies that the things humans conceive are direct extensions of ourselves; that our instrumental "handy" work produces mechanical contrivances (figure 6.4).

What would happen if the terms *organic* and *mechanic* were associated with architecture in the more intimate way suggested here? For one, the *automata* that so enticed the minds of architects, philosophers, and artists (as evidenced by Vitruvius, Aristotle, and de Chirico) might be recuperated, not as a miracle or a novelty, but now, in the twenty-first century, as a presence carrying all the accumulated connotations of the *organic* and *mechanic* that has transpired. John von Neumann, who made significant early contributions to computing among many other scientific fields, anticipated the vitality of machines in his "General and Logical Theory of Automata," in which he conceptualized the automaton less as a mechanical contrivance and more as a life-like, intelligent machine that processes and acts on information.[17] For von Neumann, life was a class of automata: understanding such life-like, human-made artifacts would reveal something significant about ourselves and the other living things that surround us.[18] Taking inspiration from von Neumann, and from the artificial life community, and also from such artistic wonders as the wind sculptures of Theo Jansen, this book strives to make the case that the term *life* should extend

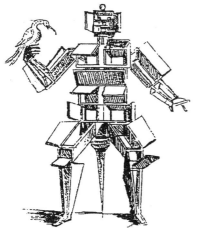 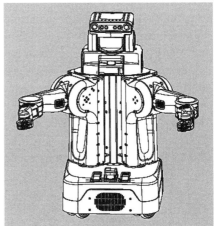

6.4

The confusion of *organa* and *machinea* persists: **(left)** furniture-man (ca. 1621–1624) and **(right)** a contemporary, humanoid robot. (Giovanni Battista Bracelli, *Bizzarria Fantastica* [drawing, ca. 1621–1624], reproduced from Alberto Sartoris, *Metafisica della architettura*, trans. T. Geltmaker and D. Ghirardo. New York: Praeger, 1972, p. 12. Drawing of the Willow Garage PR2 robot by the author.)

beyond the things of nature to include certain human-made artifacts, including those that compose segments of the built environment.[19]

Sociologist John McHale suggested something of this prospect in "Man Plus," a chapter from *Future of the Future*. Because we are so immediately familiar with our "various physical and intellectual capacities," contends McHale, we naturally "externalize [them] into autonomously evolving systems."[20] Our skin becomes clothes and houses and submarines and spacecraft. Our eyes become microscopes, x-rays, and the Hubble telescope. Our brains become information processors, computer storage, and artificial intelligence. We populate the world with visible and invisible manifestations of ourselves, and as these prosthetics, broadly defined, grow ever-more sophisticated, so argues McHale, they will increasingly augment, extend the reach of, and directly modify our human capabilities. Importantly for this consideration of architectural robotics, McHale has taken us very far from the static, lifeless world inhabited by the modular men of Vitruvius, Alberti and his contemporaries, and Le Corbusier. McHale has also taken us to a place other than that one populated by that obvious "manifestation of us" as humanoid robots. But then, what kind of life is a *living room* leading? Is it a room filled with butterflies, automatons, or something else altogether?

THE MECHANIZED HOME

What becomes of our most intimate physical environments and us as technology is introduced to the built environment is a critical question for those thinking about and developing architectural robotics. In the aftermath of the Second World War, Sigfried Giedion grappled with much the same question in *Mechanization Takes Command*: "What happens to the human setting in the presence of mechanization?"[21] Giedion did this by drawing from the history of furnishings, beginning with the furniture of the Gothic Middle Ages. Because of the profound insecurities of the Middle Ages, owners carried their furniture with them, even on shorter trips.[22] The prevalence of moving one's furniture—chests were the most common—lends to furniture the Italian word for it: *mobile*.[23] *Mobile* also has etymological roots in the High Gothic (medieval) notion that the world "was created and set in motion by an act of will," embodied in the soaring heights of the Gothic cathedral.[24] The vertically extended, magnificently slender Gothic cathedral was in stark contrast to the formal balance, equilibrium, and weightiness of classical architecture. As Giedion had it: "As the Greek temple symbolizes forces in equilibrium, in which neither verticals nor horizontals dominate, the earth in the classical view formed the forever-immovable center of the cosmos. The soaring verticals of the Gothic cathedrals mark no equilibrium of forces. They seem the symbols of everlasting change, of movement."[25]

As follows, furniture has its roots in movement: in the soaring heights of the cathedral and in the medieval master who carries his furniture to protect it from being stolen in the dark ages of insecurity.

In the Middle Ages, furniture was not modeled to conform to the shape of the human body.[26] Chairs only became commonplace after the mid-sixteenth century (that is, well into the Renaissance), and it wasn't until the eighteenth century that (mostly French, Baroque, and Rococo) furniture was diversified into types (like the writing *bureau*, for one) or molded to conform to the human body. Furniture molded to the body's form is manifested, especially, in the upholstered furniture of the *Rocaille* style, characterized by sinuous, shell-like forms of a kind that "a boneless organism surrounds itself."[27] "Rocaille," according to Giedion, "reflects the will of the period to render things flexible ..., a fantasy ruled by analysis and precise observation of human posture."[28] The physical character of Rocaille furniture is the outcome of "the upholsterer, ... destroying the structure of [conventional furniture], turn[ing] it into an invertebrate piece."[29] The sumptuously cushioned and curvy *chaise longue* is the definitive representation, which followed from a reclining daybed, the *lit de repos*, featuring a "head section ... adjusted by gears or chains."[30]

Moving across the Channel to England, furniture of the same period there was reduced to "morally pure forms" featuring, in some cases, pivoting parts.[31] The *reflecting dressing table* or *Rudd's table* (figure 6.5), referenced in the previous chapter, was of a kind selected by Giedion to represent the prevailing sensibility in English furniture of the time: essential, slender, elegant, and both mechanically and functionally flexible. What this portends for architectural robotics is that, well before the "fully-mechanized" twentieth century and the "Information World" that follows it, the earliest established blueprint for furniture clearly anticipated *living rooms* composed of refined, mobile, flexible, reconfigurable furniture, responsive to the human form in practical and sometimes fanciful ways.[32]

In *Mechanization Takes Command*, Giedion does three more things for architectural robotics. First, Giedion begins to erode the distinction between people and their physical surroundings in, for instance, defining the Rocaille as the intimate fit between furniture and its user, and by equating, throughout his history, characteristic furniture traits to prevailing personality traits. In regard to the latter, the Rudd's table follows from the "favored" human male type of its period and place ("slender, immature, ephemeral") or maybe the feminine counterpart ("mild, dreamy, ... innocent of passion"), which starkly contrasts with French Rococo furniture and its corresponding human type, reflected by "a refined, *spirituel* [sic] society, enjoying life to the point of corruptness."[33] The correspondence between people and their physical surroundings extends beyond the ergonomic,

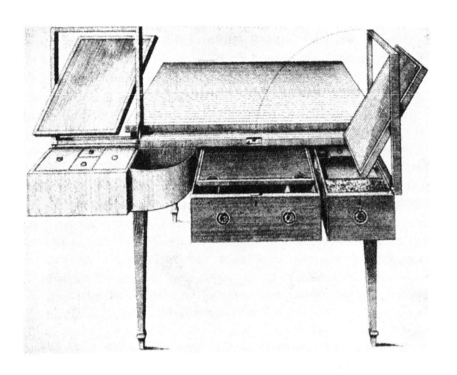

6.5

A manually reconfigurable reflecting dressing table, also known as Rudd's table (England, 1788). (Reproduced from Sigfried Giedion, *Mechanization Takes Command: A Contribution to Anonymous History*. Minneapolis, Minn.: University of Minnesota Press, 2013, plate 188 [p. 327]. Copyright 1948 by Oxford University Press. First University Press edition, 2013. Reprinted with permission of the University of Minnesota Press.)

suggesting a deeper reflectivity between our homes and us for which architectural robotics strives, as a *living room*.

The second thing Giedion does for architectural robotics is to dissolve the distinction not only between people and furniture but also between furniture and its enveloping space: the room. In *Mechanization Takes Command*, this offering comes as Giedion arrives at the twentieth century of his history, during which "the room and the contents were felt as a single entity."[34] All within the province of the architect's design, the "struts and planes" of the new furniture are light, hovering, skeletal, and natural to the point of being one with the architecture of the room that envelopes it. This present book will repeatedly show that architectural robotics is a means of breaking down the distinction between furniture and the room to the benefit of the inhabitants.

Giedion's third and final gift to architectural robotics is discovered in an early chapter of *Mechanization Takes Command* devoted to the history of physical movement, its representation, and the overall trajectory of thinking through time, collectively suggesting for environmental design "new forms, new expressive values" that should find their way as well to architectural robotics.[35] From the years following the Second World War during which *Mechanization Takes Command* was published, Giedion sent us a warning today: that, with the technological means and the creativity that abounds today, designers should recuperate the "form-giving energy" that Giedion recognized in certain periods of history, which "may perhaps be measured in comparison to our own time, with our storehouse of new materials, and our inability to give them life."[36]

This consideration of Giedion's *Mechanization Takes Command* and this chapter as a whole are concerned with the limits of architecture and computing, whereby a suite of cyber-physical furnishings can individually and collectively support multiple human needs and opportunities. For the 50th Anniversary Issue of *IEEE Spectrum*, the flagship magazine of the world's largest professional organization devoted to engineering and the applied sciences, the titles of its articles are in the form of predictions for the near future. One of these prophetic titles reads, "Tomorrow's robots will become true helpers and companions in people's homes."[37] For architectural robotics, the same title should be amended, simply by deleting a few words, to become: "Tomorrow's robots will become people's homes."

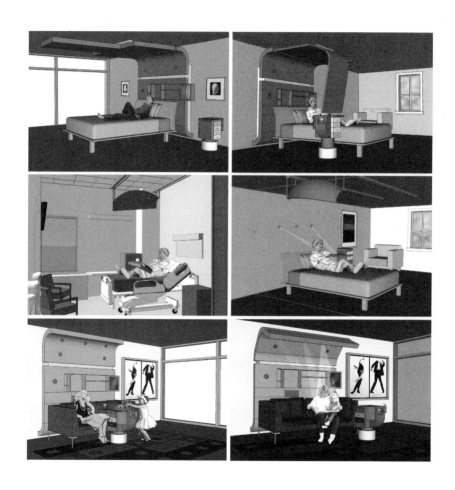

7.1

home+, in concept, at home and hospital (drawings by the author).

7 THE ART OF *HOME+*

An exemplar of the distributed environment typology of architectural robotics, *home+* is my early effort to realize the living room, to give life to the most commonplace built environment. *home+* (figure 7.1) is a suite of networked, robotic furnishings integrated into existing domestic environments (for aging in place) and health-care facilities (for clinical care). *home+* aims to augment the interior of the home or patient room to become a more responsive and accommodating environment, increasing the quality of life for people aging in place or suffering from a medical condition or disability. Very much a continuing project, *home+* strives to empower people to remain in their homes for as long as possible, even as their physical capabilities alter over time, and, in more grave circumstances, to afford people some semblance of feeling "at home," as users move between their dwellings outfitted with *home+* and assisted-care facilities equipped the same way.

This kind of assistive technology is urgently needed. While the global population continues to age, there is a smaller segment of society to both care for and pay for the well-being of older and clinical populations. It is estimated that in the United States, by 2025 there will be a shortage of physicians and nurses, and by 2040 there will be more than 79.7 million adults who will be age 65 or older—a 92 percent increase from 2011. The response from my research team, *home+*, strives to reduce the burden on health-care staff (to deliver their services) and on a decreasing tax base (to pay for this service), while meeting patient needs and ensuring the well-being of an increasingly older population.

Prior to the efforts described here, architecture and robotics have not cooperated toward realizing assistive technologies. While in hospitals, assistive technologies have become pervasive and indispensable during medical crises, and while in homes, assistive technologies proliferate as computerized health-monitoring systems and, perhaps in the future, as assistive, humanoid robots, the physical aspect of these built environments meanwhile remains largely ill adaptive to rather than accommodating to both the subtle and more dramatic life changes of its inhabitants.

Common architectural and industrial design strategies for accommodating aging and clinical populations suffer numerous shortcomings. While the conventional components (e.g., walls) of "adaptable" homes are organized to anticipate remodeling, they demand of their homeowners the tremendous will and substantial means to remodel them. And while homes and their personal effects informed by universal design principles accommodate "everyone" from the outset, they tend to suffer from the limitations of a "one-size-fits-all" misconception.

Common computing strategies for accommodating aging and clinical populations suffer their own shortcomings. While homes and assistive residences outfitted with ubiquitous computing technologies (e.g., sensor networks, camera networks, and RFID tagging) can detect crisis, can support tasks such as taking medication, and promise peace of mind to caregivers, such a "smart home" approach has a range of difficulties: its technology is distributed most everywhere, whether desired or useful; its inhabitants intervene with (i.e., break) the technology; and its systems suffer from accidental mishaps and are otherwise not sufficiently sensitive to individual differences and desires (in particular, the demand for privacy). As for the possibility of an in-home, humanoid service robot, data suggest that while people may welcome a service robot to compensate for their reduced capacities, they don't care much whether the robot looks or acts particularly human.[1] Previous robotics research for health-care and elder-care applications have tended to focus on specialized devices aimed at dedicated tasks, such as rehabilitative robotics, robot-assisted surgery, and prosthetics. Closer to our vision of assistive domestic environments embedded with robotics is the concept of the "Intelligent Sweet Home" from KAIST (the Korea Advanced Institute of Science and Technology); however, in this research effort, the only robotic component receiving significant attention, to our knowledge, is a hoist for transferring users to and from their beds—a robotic device serving a singular, practical function.[2]

The design challenge of our expanded design-research team, which included human factors psychologists and domain-specific health-care providers, was to understand the target populations (both patient and provider) and design an assistive, technological system for and with them. This challenge required a creative approach to research involving both design and engineering innovation and concentrating "on functionality, aesthetics, and manufacturability, simultaneously."[3] In other words, this was another underconstrained, underdetermined, wicked problem for architectural robotics, much as was the workplace for colocated information workers that resulted in the Animated Work Environment (AWE).

BIG VISION BEFORE BIG DATA

Given the complexity of our ambition at room scale, my research team and I elected to focus initially on developing the larger *home+* vision of an assistive, architectural robotic environment composed of a number of partly intelligent, interactive furnishings.[4] Our *big vision* approach is contrary to the typical approach of creative design practitioners and robotics researchers, whose attention tends to be focused on the design of the intended artifact as primarily an isolated one. Instead, in conceptualizing *home+*, we quickly prototyped our larger, architectural robotic concept as an integrated vision in all its complexity: the design of various furnishings interacting with each other and their users in a specific built environment (existing room). This activity is analogous to that for AWE, in which we quickly developed a full-scale prototype before we had acquired sufficient data to guide it responsibly in a measured, industrial design and engineering sense. The virtue of our quick and early, big vision prototyping activity permitted us to make inevitable human mistakes faster. Such big vision prototyping attends to the expansive territory that urban designers call their home: their system of relationships (their "Internet of Things").

In our early, focused effort, we envisioned *home+* as having five key components (figure 7.2):

- An *assistive robotic table* gently folds, extends, and reconfigures to support therapy, work, and leisure activities.
- An *intelligent headboard* adapts to the profile of the patient's back, morphing its supportive surface periodically to alter the patient's position as a vehicle to ensure against bedsores. The intelligent headboard also offers storage, accommodates an oxygen tank as required, and provides intelligent lighting for reading, work activities, and ambient illumination.
- A *sensitive mobile rail* is not unlike the vertical tubular-metal post found in metro cars, except that this one moves with a user, providing more or less assistance, like the arm of a friend, as the user moves between one location within the dwelling to another location, tracing the most common paths traveled in the course of a day. As envisioned, the sensitive mobile rail recognizes how much assistance a user might need, and "backs off" from supporting the user when it senses the user is managing fine with less of its help.

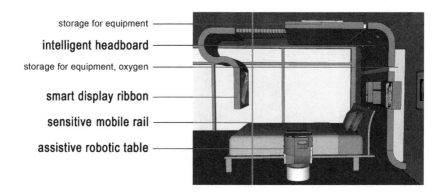

7.2
home+ components (drawing by the author).

- *Intelligent storage* (figure 7.3) manages, stores, and delivers personal effects (including medical supplies) and communicates to caretakers when eyeglasses and other belongings are not moved over a period of time.
- A *personal assistant* (not shown in figures 7.2 and 7.3) retrieves objects stored around the room and away from the bed. The robot uses a vision-based recognition system via wireless communication to ensure the robot retrieves the correct item.

These five components of the *home+* suite recognize and communicate with each other in their interactions with humans and, we envision, with still other *home+* components. Figures 7.1–7.3 provide an impression of how the system might look and behave, not its fixed, predetermined design. Indeed, toward realizing the larger distributed system of robotic furnishings, the research team persists in continuing iterative design and evaluation of the *home+* concept, generating alternative, conceptual visualizations of *home+*, and developing numerous physical, functioning prototypes of its individual components, identified here and otherwise. In addition to developing some manifestation of the above *home+* components, we have also developed, to varying extents, the following additional *home+* components: a smart mattress, a smart sofa, a smart and mobile side table, smart lamps, a physically reconfigurable lounge chair, a smart articulating arm for mounting computer displays and televisions, and a nonverbal communication system for interacting with *home+* (the latter elaborated in this chapter).

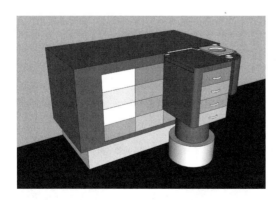

7.3
home+ intelligent storage with table, docked (drawing by the author).

THE ASSISTIVE ROBOTIC TABLE

As the research team wished to focus its effort not only on the big vision of the *home+* concept but also on a more developed and refined component of it, we dedicated considerable efforts on the *assistive robotic table* (ART), given its relative complexity within *home+*. Our aim was to develop this key component as a fully functional prototype with the participation of clinical staff.

ART (figure 7.4) is a hybrid of the typical nightstand found in homes and the over-the-bed table found in hospital rooms. We envision ART integrated into the domestic routines of its users, even as users transition from home to clinic and, hopefully, home again. Our research team hypothesizes that individuals using ART as part of their domestic landscape will live independently longer. Moreover, we expect ART to free familial caregivers from performing certain arduous tasks for ART's target populations, allowing caregivers to devote more energies to meaningful human interaction with ART's users. In medical facilities, ART aims to augment the rehabilitation environment by improving patient well-being, rehabilitation, and staff productivity (in this vexing moment of limited resources). As already stated, ART and the other components of the *home+* vision recognize, communicate with, and partly remember each other in interactions with human users.

Along with the larger *home+* environment to which it belongs, ART benefits from the convergence of advanced architectural design, computing, and robotics largely absent from prior efforts in assistive technologies. In particular, this enabling technology is not distributed everywhere in the physical environment but *where it's needed*; it is not intended to be invisible but *visible, attractive, and integral* to the home and the patient room by design; and it is not meant to serve as a means for surveillance but rather as *environmental support* that recognizes and dignifies what people can do for themselves.

7.4
The assistive robotic table (ART)—our final, fully functioning prototype.

These attributes of ART and the larger *home+* environment are of a kind identified by Donald Norman as "the next UI [user interface] breakthrough," defined by "physicality," and accomplished with "microprocessors, motors, actuators, and a rich assortment of sensors, transducers, and communication devices."[5] In broad theoretical terms, Nicholas Negroponte anticipated ART and the larger *home+* environment in the 1970s in his vision of a "domestic ecosystem" that regulates aspects of "environmental comfort and medical care."[6] More recent inspirations for ART include Malcolm McCullough's plea for "architecturally situated interactions," which "permit the elderly to 'age in place' in their own homes."[7]

KEY COMPONENTS OF ART

Physically, ART is a significant development of the over-the-bed table commonplace in hospital patient rooms. What distinguishes ART from the conventional over-the-bed table is its novel integration of physical design and functioning, coupled with an interactive human-object interface (figure 7.5). Integral to ART is a novel, plug-in, continuum-robotic *therapy surface* that helps patients perform upper-extremity therapy exercises of the wrist and hand, with or without the presence of the clinician (see figures 7.8 and 7.13 later in text). Considered in depth in a subsequent section of this chapter, this therapy surface was recognized as a key requirement for ART, following our early conceptualization of the larger *home+* vision, and as a significant outcome of our ethnographic investigations in the clinical setting (or in the "wild," as the field of human-computer interaction [HCI] defines territories outside the lab). These ethnographic investigations suggested the promise of the therapy surface for rehabilitating, in particular, the upper extremities (upper limbs) of post-stroke patients as will be elaborated shortly.

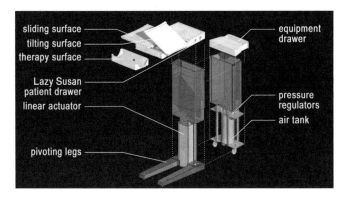

7.5
ART's key components.

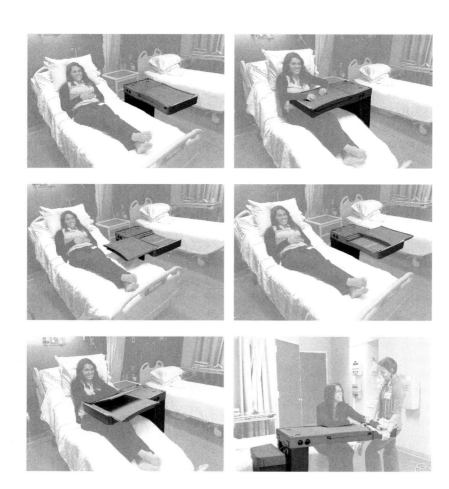

7.6
In the hospital with ART and "Amy": six instances.

Scenario 1
AMY MEETS ART FOR THE FIRST TIME
Amy had a stroke and has right hemiplegia (paralysis of the right extremities) and aphasia (speech and language problems). After treatment in the hospital, Amy returned home, fitted with *home+*, the same suite of robotic-embedded furnishings that supported her in her patient room. *home+* has uploaded Amy's preferences as acquired from her hospital stay and modifies these preferences and those of her caregivers over time to best support Amy's recovery. Amy relies on the assistive robotic table of *home+* every day (figure 7.6): the therapy surface of ART helps Amy rehabilitate her arm; ART tilts and changes height to best accommodate Amy's activities; ART's nonverbal lighting cues remind Amy to take her medications; and ART learns and adapts to the gestures Amy performs with her arm as she regains more capacity in moving it. ART logs Amy's reading time as a wellness metric, and ART initiates storage of Amy's reading glasses when she's finished reading. These functions and others help Amy improve more quickly. ART's components recognize and partly remember, communicate with, and cooperate with her, her caregivers, and the other components of *home+*, empowering Amy to remain in her home for as long as possible, even as her physical capabilities alter over time; and, in more grave circumstances, affording Amy some semblance of feeling "at home" as she moves to an assisted facility also equipped with *home+*.

Scenario 2
HOW AMY ADDRESSES ART, AND HOW ART LEARNS ABOUT AMY
As an early design exemplar of the distributed environment typology of architectural robotics, by which furnishings seemingly, collectively, *come to life*, ART and the larger *home+* suite occupy a peculiar place between the animate and inanimate. If ART has something of a lively presence—more so than a beloved car or living room chair, less so (maybe) than a humanoid robot—how should you best address ART? How do you converse with an animated, architectural robotic environment?

The scenario just presented, featuring Amy and ART, didn't offer a reply to this critical aspect of human-machine interaction. As a way of remedying the matter, what follows is the next episode of Amy and ART's story.

Amy, a post-stroke patient, is getting acquainted with ART while rehabilitating at the rehabilitation hospital. In the later afternoon, Amy's sister, Joanne, enters Amy's patient room. Amy asks Joanne to borrow her laptop computer in order to check her e-mail. To accommodate the computer, Amy would like ART to raise and position its flip tray for her. Amy still feels a little unsteady holding everyday objects, and ART can provide the needed support for this activity. While Amy and Joanne continue their exchange, the following nonverbal dialogue occurs between Amy and ART (figure 7.7).

Amy: [Gestures toward ART to rise and tilt, as if to say, "ART, please rise and tilt for me."]

ART: [Displays two quick light flashes and emits a beep-beep sound as if to say, "Yes, I am pleased to do this for you."]

All the while, Amy and Joanne are chatting, catching up on recent news in their lives and the world. A few moments later:

Amy: [Gestures for ART to raise, but ART does not comprehend at first.]

ART: [Displays blinking lights and emits a sound that, if written, might sound like *ond-ond,* as if to say, "Hmmm, I don't know what you are asking of me. I am puzzled."]

Amy: [Makes the gesture once more in a way that ART comprehends, learns from, and responds to with the correct behavior.]

To reinforce ART's actions:

Amy: [Runs her hand along ART's sensors at the perimeter of the table, in what appears to be a "pet" to convey to ART, "Thank you."]

ART: [Displays a gradient on/off light pattern, as if to say, "I understand that I performed the task correctly."]

7.7
The nonverbal communication loop: "Amy" and ART communicating through gestures, lights, and sounds.

In this nonverbal dialogue between Amy and ART, Amy gestures toward ART. In most cases, ART recognizes Amy's arm-scaled gestures and accordingly responds to Amy with its assistive support, accompanied by a gentle sequence of lights and sounds as a means for providing further feedback to Amy. In the more atypical case (as offered in the scenario) where ART doesn't recognize Amy's gesture and so fails to understand her request, ART emits a different sequence of lights and sounds to communicate ART's failure at understanding Amy. Accordingly (as offered in the scenario), Amy can repeat her gesture for ART now that ART is prepared to read Amy's request with "greater attention" so that the system better understands Amy the next time Amy makes the same gesture. Amy then rewards ART with a "pet" so that ART's behavior is reinforced. When the exchange between Amy and ART is easy, not rife with misunderstandings, then this kind of nonverbal, human-machine exchange between Amy and ART prevents, at least partly, any disruption to the flow of Amy and Joanne's intimate (human-human) conversation. If the exchange between Amy and ART was verbal, not nonverbal, then obviously the words exchanged between human and machine would arguably disrupt the flow of communication between Amy and her sister.

Our team has published results on the nonverbal exchange offered by ART, both its capacity to learn gestures from its human users and its ability to emit sounds and words as a means of response. However, we do not yet have for ART the real-time implementation of this nonverbal communication loop, nor are we yet convinced that nonverbal communication is the most appropriate means of human-machine exchange for architectural robotic components and environments.[8] Nevertheless, users don't expect a piece of furniture or a physical environment (like a room) to speak to them in spoken words, much as they would expect of a humanoid robot that looks and behaves, at least vaguely, like a human being. This implies that lights and sounds may not be a bad communicative vehicle for architectural robotics. Moreover, the vocabulary (verbal or nonverbal) that users might expect from one piece or a suite of interactive furniture is far less than the vocabulary expected of a humanoid robot that, again, given its human-ness, suggests to users that it's capable of higher-level intelligence. This implies that if we envision new users experiencing difficultly in deciphering ART's nonverbal language of lights and sounds, we can take some assurance from the fact that ART, as a piece of furniture, doesn't suggest to users that it is capable of accomplishing much more than a small part of what a humanoid robot promises to do on the basis of appearances. And ultimately for architectural robotics, a component such as ART and an environment such as *home+* might do enough for us. Do we need a very human, humanoid robot serving us? And while we wait for the realization of a very human, humanoid robot (as one doesn't exist yet), how long can we endure living with a dumber one, and wouldn't we prefer to have smart furniture?

We cannot yet answer these questions in the way they deserve to be answered: by way of longitudinal studies of people living with advanced exemplars of architectural robotics and humanoid robots (or, someday, combinations of both).

We will nevertheless return to these questions, at least to ponder them, in the course of this book. For now, returning specifically to the matter of human-robot communication (irrespective of the robot's physical form or intelligence level), a number of research studies have suggested that people tend to react adversely to robots exchanging in and (particularly) issuing commands in spoken language: these studies suggest, instead, that people are relatively more receptive to nonverbal communication emanating from robots.[9] Indeed, it has been shown that people can easily interpret the meaning of nonverbal utterances presented by robots.[10] As well, people who are ill or in pain tend to reduce their level of verbal communication, making relatively more use of nonverbal communication compared to when they are well.[11] Furthermore, the nonverbal communication of American Sign Language has been shown to be more effective "than spoken English because of the linearity of spoken language."[12] While these results and observations are hardly conclusive, collectively they underscore the validity of a nonverbal approach like ours to architectural robotics and, more broadly, to human-robot interaction for robots of wide-ranging appearances and behaviors.

In designing our nonverbal communication loop for ART, we codesigned, with clinical staff, ART's audiovisual system, defining its technical means, its physical location on the artifact, and its vocabulary of light and sounds to provide feedback to users. We also designed, mostly as a working concept that we are only now implementing in the artifact, ART's capacity to "learn and recall" the arm-scaled gestures of users, some of whom we expect would not be capable of producing with consistency all of the gestures that might be hard-programmed into such an assistive device. We conducted early evaluations within the real-world situation of a rehabilitation hospital where the stakeholders—clinicians, post-stroke patients, and their intimates—can help us advance our developing nonverbal communication loop. Research on nonverbal human-machine communication occurring in situ—in the intended setting for its usage—is relatively scarce. Lamentably then, what Justine Cassell, professor of HCI at Carnegie Mellon University, offered some years ago remains true today: "there has been little research" situated "in the wild" and "focused on bi-directional human-robot communication employing models of nonverbal communications as both input and output."[13]

Clearly, more work is needed in this domain, as assistive and other human-centered cyber-physical technologies increasingly take different forms and exhibit different levels of intelligence. Voice recognition, for one, may not be the only means for communicating with all of these artifacts. Our technical paper on the iterative process of designing and evaluating ART's audiovisual system suggests the promise of this communication strategy for some kinds of assistive robots.[14] For the purposes of defining architectural robotics more broadly, however, a brief consideration of ART's capacity to learn and recall gestures, its gesture-recognition system, captures the important question of intelligence in architectural robotics. Recall Negroponte's prescription that, for intelligent environments, "some 'learning' is involved" to render them "purposeful if not [strictly] intelligent."[15] Accordingly, architectural robotics exhibits a little intelligence, as suits the purpose; and achieving a little intelligence is indeed a more modest ambition than the strivings of certain members of the HRI community in designing highly intelligent humanoid robots, from which my lab and I nevertheless take inspiration. Compared to researchers in humanoid robotics, however, my collaborators and I envision the components and environments of architectural robotics—ART and *home+* being early exemplars—relying far less on their own machine intelligence and more on the intelligence *of their users* to operate. In this way, the components and environments of architectural robotics work *with* rather than *for* the user. They work collaboratively, in tandem.

With respect to gesture recognition, the prevailing model of it works against our vision, requiring users to adhere to its self-centric, programmed understanding of the world. To be more specific, the prevailing model of gesture recognition relies on template matching, in which the human user is expected to emulate a choreographed motion as prescribed and programmed by the researchers. This means that users, wishing to communicate with assistive cyber-physical artifacts or robots, must submit to them, to their programming, ensuring that the gesture maps users perform match precisely the template classification found in the machine's programmed library of distinct gestures.

In our early effort to design and implement nonverbal communication for architectural robotics, we designed for ART a system that recognizes the unique and changing physical and mental capacities of users and their individual ways of expressing themselves, even through physical gestures produced with the arm.[16] In more precise terms, we designed for ART a gesture-recognition system that, first, recognizes a range of gesture performances by users, given that their mental and physical capacities are unique to them and also may vary dramatically day to day as their medical conditions improve or worsen; and second, is capable of learning new gestures as defined by the users themselves, as each user may have a preferred way of communicating.

Given the prevalence of arm-scaled gesture as an everyday means of human communication, our system promises an interface that is effective, extensible, and intuitive to the user.[17] Because of the range of possible performance variations by the user (who may have suffered a stroke, we assumed in this investigation), the ability of such an interface to generalize across gestures quickly (i.e., following the fewest observations made by the machine) is essential to its success. Also essential to the system's success is its ability to learn new gestures with no adherence to a choreographed example, so that users who suffer impairments or who are unfamiliar with the system can benefit from its assistance with less effort. Addressing these challenges are the key contributions of our research.

Designing an automated gesture-recognition system for ART and other assistive technologies involves the combination of a sensor platform, data representation, pattern recognition, and machine learning. The sensor platform is used to capture the gestures of the user. For ART, we used the RGB-D depth sensor of the Microsoft Kinect, which provides a rich, real-time, three-dimensional (3-D) data stream that preserves user anonymity and functions in dark environments. Compared with the depth-sensing Kinect, video cameras commonly used for the same purposes have a number of drawbacks, including their tendency to work ineffectively in dark environments, their difficulty with occlusions (where, from the camera's perspective, one object is hiding another object behind it), and most significantly for us (from our user-centric perspective) their more obvious intrusion into users' privacy (even if they are not capturing the sound and moving image of users in detail, users nevertheless perceive this invasion when a camera is simply, visibly, present).[18] For the experiments conducted with ART, the camera within the Kinect was disabled and hidden to ensure anonymity.

Data representation is the visual form of the user's gestures that ART can recognize and decipher. Unlike a video camera that captures the users' gestures in two dimensions, the Kinect depth sensor extracts (i.e., simplifies) users' gestures as a 3-D data stream of vectors. This 3-D data stream is also, conveniently, view-invariant, meaning that the foreshortening of a user's limb while performing a given gesture does not result in a recognition problem for the recognition system.

After the Kinect senses a user's gesture as a 3-D data stream of motion vectors recognizable to ART, ART needs to understand what it's looking at so it can perform its response. And this kind of perceiving for ART is fundamentally a lot like perceiving for us: it's about pattern recognition. (Recall our previous, related considerations of Alexander's *A Pattern Language* and the user patterns of the AWE for supporting working life.) What ART needs to do, then, is to classify the 3-D data stream of motion vectors, representing the user's gesture, into one of the known gallery types recognized by the recognition system. Researchers in human-robot interaction often use *hidden Markov models* (HMMs) for classifying gestures, an approach that intuits the most likely (yet "hidden") trajectory for the gesture type from a sequence of visible observations of the gesture performed by the user. However, HMMs are not the ideal model for ART's recognition system, for two reasons: first, for HMMs, translating visible data is a partly inaccurate process, which makes it somewhat insufficient for ART and *home+*; and second, and more critically, HMMs require that what the system "sees" is predetermined, whereas we are assuming with ART that *users* will determine what the system "sees" (and the system will, accordingly, learn and adapt to the manner in which individual users gesture).[19] More apt than HMMs for our bidirectional, user-centric vision of ART's gesture-recognition system is a recognition scheme based on the growing neural gas (GNG) algorithm. Unlike HMMs, a GNG algorithm places no initial constraints on users to perform gestures in any specific way. Our results to date show that the GNG algorithm allows for generalization across gestured commands, enabling (in concept) ART to understand users' gestures even when users don't or can't perform them precisely. This capability is much like intelligent recognition systems for handwriting and voice that recognize handwritten characters and spoken words for most any user, despite the unique ways in which individuals express themselves. This attribute of ART should prove particularly comforting to new users and users faced with physical or mental challenges, as the recognition system is aimed at recognizing their idiosyncrasies and making sense of them.

We thus far, in tracing ART's process of gesture recognition, have ART recognizing the gestures of users as they produce them. And once ART recognizes the particular pattern gestured by the user (e.g., the gesture that means, *ART, tilt your work surface*), ART maps the given pattern directly to a robotic response (and ART's surface is tilted). In this process, however, ART has not yet been rewarded for its hard work, nor has ART learned any new tricks (new gestures). While ART's capacity to generalize gesture commands may be considered an instance of machine learning, our use of the term applies to a mechanism by which users provide some manner of feedback to improve future outcomes of ART's assistive behavior. This feedback may take the form of a simple good/bad reward signal provided by the user, as employed in ART. Typically, training of a system is accomplished "off-line" (prior to in situ, in-the-wild use of the assistive system). Maybe you remember how this worked with early versions of handwriting-recognition systems, like Newton, Apple's personal digital assistant of the early 1980s. Newton was typical of such devices, requiring the laborious task of writing characters repeatedly on the tablet as a means of training the system to recognize your handwriting in preparation for its first "real" recognition activity. For these machine-learning systems, a large number of iterations during the training phase are required before it can be put to use. Our objective for ART was, instead, to create an online learning modality that utilizes direct interaction with a user to enable ART to converge upon the particular response desired by the user. In short, our aim for ART was to create a direct mapping between sensed gestures and inferred goals. Generally, such a convergence between sensed gestures and robotic responses is impractical to achieve, given a number of factors: the potentially large number of pairings between sensed gesture and actuated response (*state-action couplets*); the mechanical limitations of execution speed; issues of reliability; and energy consumption. This is where the rewards become important: if the user provides a simple reinforcement of ART's behavior—an indication that the response was either good or bad—the system learns more quickly. This reward can take the form of a simple mouse click or push of a button; but in the case of ART, ART would much prefer the gentler, more respectful pet or stroke.

Overcoming many of the shortcomings of previous efforts in gesture recognition, our early investigations of ART's gesture-recognition system show that 3-D data from an RGB-D camera can be used to generate a useful descriptor of gesture; that the GNG algorithm is capable of differentiating between these descriptors more effectively than a conventional approach; and that learning which typically requires a large number of iterations to produce effective results can be accomplished more quickly with a small data set when employing our GNG and machine-learning approaches. There is plenty more for us to do before gaining confidence in ART's capacity to decipher and learn gestures performed by users in ways that can satisfy our user-centric objectives. Nevertheless, we believe there is promise in our approach to gesture recognition as a mode for interactions between architectural robotic components and environments and their users of wide-ranging capacities, dispositions, and interests.

In tracing the workings of our nonverbal communication loop, we arrive, once again, at the matter of pattern recognition, having considered pattern recognition in more theoretical terms previously. Specifically, in chapter 4, we explored pattern recognition from different perspectives (Gestalt psychology, Alexander's *A Pattern Language*, computational intelligence), arriving at the understanding that we human beings structure our experience of the world by way of patterns. Meanwhile, ART's nonverbal communication loop is, really, far less than a loop, as the machine in this case of human-machine communication exhibits intelligence far inferior to our own. Consequently, ART's human-machine communication loop, composed of gestures, lights, and sounds, is more a deformed or lopsided loop (a pear?). What accounts for the lopsidedness is the difficulty machines have, at least currently, in recognizing and learning a wide-ranging *command vocabulary* of patterns—something that comes naturally to us. In the end, *we* are the pattern-recognition machines, and ART and its cousins, even the humanoids, are lower-level relations.

BUT WHAT KIND OF INTELLIGENCE?

As an intelligent machine, ART operates at a level that lies somewhere on the continuum between an application-specific robot (a Roomba vacuum) and a humanoid robot (your personal assistant). Given that ART performs not one assistive function but a number of assistive functions, and given that ART is capable of learning, it leans closer, as an assistive robot, to the more capable, humanoid end of the continuum. But why is this location on the continuum the right place for architectural robotics? Remember again, with respect to intelligent environments, Negroponte's wise prescription: that the right measure of intelligence is that amount that suits the artifact's purpose. And, as Negroponte concluded, that amount of intelligence is not likely much.

Negroponte's prescription dates to 1975 with the publication of *Soft Architecture Machine*, introduced in chapter 2 with Alexander's *A Pattern Language*. Ten years earlier, Moore's law predicted that computing capability, measured as the number of transistors on an integrated circuit, would double every two years. Computing capability therefore increases geometrically, which has since proved true. And this fact has in part triggered an obsession with machine intelligence among some of today's researchers, tech leaders, and tech chroniclers: How powerful can we make artificial intelligence?, How fast can we create it? When will machines match our human intelligence and even surpass it?

Most visibly entangled with these questions is the tech leader and chronicler Ray Kurzweil, whose website bears the title, "ACCELERATING INTELLIGENCE." In Kurzweil's book of 2005, *The Singularity Is Near: When Humans Transcend Biology*, he confidently offered 2045 as the arrival of the *singularity*—the year in which machine intelligence is expected to surpass human intelligence to the extent that it alters the core of human existence.[20] In any extended discussion or consideration of the singularity, there is inevitably the proposition that human beings can transcend death by uploading their inner selves to machines, as machines, some argue, will have become intelligent enough to bear us. Various esteemed voices have debated the likelihood of this kind of high-tech immortality and, generally, the singularity. Most prominently, the debate unfolded in a special report of *IEEE Spectrum* (the magazine of the Institute of Electrical and Electronics Engineers referenced earlier) and continues in its associated webpages and blog.[21]

Among the prominent voices included in the IEEE special report, robotics pioneer Rodney Brooks argues, from the optimistic side, that "machines will become gradually more intelligent ... with cognition and consciousness recognizably similar to our own"; but "not by the imperative of the singularity itself but by the usual economic and sociological forces."[22] Brooks treads a bit carefully here, not exactly rejecting the singularity (i.e., "not by the ... singularity alone"), nor stating that cognition and consciousness will *match* our own, but instead will be *recognized* as such by us.

In any case, perhaps we can imagine another singularity, one not so faithful to the many voices participating in the IEEE debate: a singularity between the natural environment and the intelligent environment rather than between human intelligence and machine intelligence. A singularity between the natural environment and the intelligent environment, in other words, is the aspiration to conceive an artificial environment that can match the capacity of the natural environment. No small feat. However, achieving the singularity of the IEEE debate is no small feat either, yet it is certainly advancing in machine learning and other corners of software and in hardware in the arena of humanoid and biomimetic robotics. One might add to this list the corresponding advances in digital animation, particularly the capacity to emulate, rather faithfully, complex behaviors found in nature, such as the movement of hair and muscle under different conditions. But despite these advances toward the singularity (however broadly defined), this effort is in its infancy with respect to a built environment aspiring to Negroponte's promise of an "artificial domestic [eco]system," introduced earlier in chapter 2. As offered throughout this book, the singularity of the natural and the intelligent environments is an apt conceptual framework for supporting and augmenting a range of human activities for an increasingly digital society. Indeed, while Rodney Brooks envisions the singularity permeating not only our "bodies" but also "*our environments*," researchers in human-robot interaction and generally in human-machine systems have been, for the most part, almost myopically attending to realizing this ambition for humanoid and other animal-like "bodies" at the neglect of our built "environments." Architectural robotics seeks to remedy this by developing novel and new robotic components and assemblages destined for everyday spaces.

But how well does intelligence operate in such mundane, everyday places? What measure of intelligence should architectural robotics exhibit at home, in the office, in the hospital room, or in the schoolroom? We have Negroponte's answer from the mid-1970s, but let's spend some more time here pondering the critical question of intelligence for architectural robotics. Toward an apt reply, Jeff Hawkins, neuroscientist and founder of Palm Computing, references the subject of pattern that was considered earlier in this chapter, offering a very concrete example in a very familiar place: the typical entry door found at the foot of the typical house. As Hawkins offers, imagine that you come home one evening to discover that something, one thing, has changed in the appearance of your front door: one of a "thousand changes that could be made to your door, unbeknownst to you."[23] Maybe your front door is wider, or longer, or lighter, or a different color, or its knob has been repositioned, or its hinges have been reversed, and so forth. Hawkins points out that while you'll easily be able to recognize the difference in your door—that single change in your door across all possible changes that could be made to it—the computing approach to the same recognition problem, that of artificial intelligence (AI), would be "to create a list of all the door's properties and put them in a database, with fields for every attribute a door can have and specific entities for your particular door. When you approach the door, the computer would query the entire database, looking at" every and all properties of your door and all doors, and compare these properties to the properties of the door it "sees."[24] In this example, the difference between human and machine capacity is real and far-reaching. As Hawkins concludes:

> The AI strategy is implausible. First, it is impossible to specify in advance every attribute a door can have. The list is potentially endless. Second, we would need to have very similar lists for every object we encounter every second of our lives. Third, nothing we know about brains and neurons suggests that this is how they work; and finally, neurons are just too slow to implement computer styled databases.[25]

Implicit in Hawkins's concrete example and conclusion is that computers and the built environments are both low-intelligent in their own lowly ways: computers can't easily infer a single change to a common door from among all possible changes in appearance that a common door might exhibit, and doors can't change their appearance by themselves—they need us to do it for them.

In keeping with the same concrete example, what architectural robotics promises to accomplish is a door that can change an aspect of its appearance *by itself* in a way that is recognizable not only to us but also to other parts of the house in a purposeful way, benefiting the inhabitants and the larger ecosystem of bits, bytes, and biology. In the process of designing such an environment, architects bring an understanding of the ways in which the door can and should change in appearance, and the ways in which building typology and environmental cues, as McCullough suggests in *Digital Ground*, can help the larger system of components recognize each other, thereby compensating for AI's current shortcomings.[26] On the computing side, implicit in architectural robotics is a newer perspective on intelligence that, as characterized by Hawkins, doesn't "possess anything like human intelligence and certainly couldn't pass a Turing test," but instead "uses machine learning, massive data sets, sophisticated, and clever algorithms to master discrete tasks."[27] Here is architectural robotics manifested as *machine-environment*, "built to accommodate specific tasks in ways that people never could."[28] In the larger *home+* suite, of which ART is a key part, we recognize the promise of this kind of intelligence in a networked suite of interactive furnishings having a small degree of intelligence, as suits their individual and collective purposes, functioning together to assist users in their everyday routines and environments.

ART IN THE HOSPITAL, UNDERGOING ITS OWN EVALUATION

While we may understand something about ART's intelligence now, we also know from life experience that intelligence doesn't always translate well in navigating the real world. Accordingly, in developing the full-scale, full-functioning ART prototype, we needed to subject ART to the clinical environment where health-care assistance is delivered and received (figure 7.8). In so doing, my research team and I conducted five iterative phases of research.[29]

ART was designed to assist individuals aging in place at home and patients receiving care at health-care facilities. In developing ART in situ, we therefore had a choice of working, initially, with a population aging in place in their homes or a clinical population at the hospital. Early on in the investigation, my research team made a commitment to developing ART for the demands of critical care: to accept the challenge of creating a key work of interactive furniture for this most difficult context. Consequently, we involved physicians and occupational, speech, and physical therapists in the five iterative phases of our research. We conducted these research activities within our purpose-built *home+* lab at the Roger C. Peace Rehabilitation Hospital of the Greenville Health System (Greenville, South Carolina). As patient populations of rehabilitation hospitals typically have a high number of post-stroke patients, and as such patients partake in therapies that use over-the-bed tables, the post-stroke population was a particularly apt target for our research on the design of a forward-looking, over-the-bed table.

7.8
ART in the hospital, undergoing evaluation.

In retrospect, having performed the five phases of research in the clinical setting, we now believe that ART might have proved a greater success had we confined its iterative evaluations to the domestic environment, supporting users aging in place, rather than subjecting ART to the clinical setting, inhabited by hospital patients exhibiting, often, both cognitive and physical deficits. Consequently, we intend to subject ART to longitudinal research studies within a fully functional *home+* suite, installed in domestic environments that exist outside the significant constraints and complexities of clinical settings.

It should be noted from the outset that the larger arc of ART research has yet to involve direct participation by post-stroke patients themselves but has involved instead their clinical caregivers, playing the role of their patients as well as assuming their own role as health-care providers, in the codesign of ART. Consequently, "patients" are here more broadly defined as *personas*; that is, fictional characters representing different user types likely to interact with the artifact being developed.[30] The use of personas is commonplace in design research across the allied communities of HCI, *user-centered design*, *interaction design*, and *user-experience design*. In the case of ART, post-stroke patients are suffering too many deficits and recovery challenges to interact with such a complex, cyber-physical machine of considerable size and weight in these relatively early stages of development in a long course to full implementation. Consequently, the health-care providers, having the theoretical and practical understanding of post-stroke patients, assumed the role of personas representing the breadth of this population (table 7.1).

Patient Persona	Description
Low-functioning patient	Ted is a 71-year-old male with hypertension, admitted one week ago after suffering a severe ischemic stroke. Ted has no movement in his left arm, and he has "tunnel vision."
Medium-functioning patient	Ginny is a 64-year-old female with diabetes, admitted two weeks ago after suffering an ischemic stroke. She has no fine motor control in her right arm, and she forgets recent events.
High-functioning patient	Bob is a 52-year-old male with a family history of hypertension, admitted one week ago after suffering a mild ischemic stroke. He lacks full fine-motor control.

Table 7.1
Patient personas representing three levels of functioning in patients, as played by clinicians in phase 1

Generally, design researchers developing architectural robotics artifacts for use by patients or by other vulnerable populations must proceed with care in their research activities and must secure necessary approvals and invite monitoring (by, for example, institutional review boards). As well, it is not an uncommon condition for conducting research in the health-care domain that demographic data for participants must be withheld from all published reporting of the research activities and their findings. In the case of ART, only now that we have developed the fully functioning prototype (as shown in figure 7.4) do we have the confidence to invite patients with various physical and cognitive challenges to serve as participants in future evaluation activities. This future research would ensure that ART accommodates the wide-ranging needs of the target population. To arrive at this place, however, required the completion of five phases of research described in brief below: results from each of the five phases were subjected to team discussion and helped the team form a comprehensive action plan that promised to further the design in the subsequent research phase.

Phase 1: Needs Assessment of the Clinical Space and Technology
During phase 1, the research team observed how occupational, speech, and physical therapists work with patients, and how the clinicians used the current over-the-bed table in their acute care setting. After our observation task, the research team consequently conducted structured interviews with role-playing clinicians (table 7.1) in a mock-up hospital room to determine "how and why" over-the-bed tables are used in clinical caregiving tasks primarily by occupational and speech therapists. Clinicians reported that over-the-bed tables used in clinical settings are an issue for patients and clinicians alike: over-the-bed tables are difficult to maneuver and prone to break. Nevertheless, clinicians reported that over-the-bed tables are essential to patients during mealtimes and when patients, alone in their rooms, need to access and store items close to their beds.

Phase 2: Confirmation of Clinicians' Needs
In phase 2, the research team aimed to identify and understand three things about over-the-bed tables: the clinician's view on issues related to over-the-bed tables; the research team's view on what constitutes a better designed, interactive over-the-bed table (e.g., an assistive robotic table); and compatibility between the view of the clinicians and the view of the research team. Clinicians and the research team strongly agreed on three requirements for ART:

- All unit controls of ART must be easy to use for one person who is using one hand of limited dexterity.

- ART must be capable of bracing a patient's weak arm during therapy-strengthening of the strong arm and core muscles.
- The therapy surface for ART must provide programmable visual cues to stimulate awareness on the neglected side of patients suffering hemiplegia.

Additionally in phase 2, the research team observed occupational therapists in the inpatient and outpatient rehabilitation settings. The team observed how occupational therapists interact with patients, identified the therapy tools used primarily for fine motor control by therapists, and determined measurements for patient improvement. All this information helped the research team identify aspects of an assistive robotic table that promised to be successful in clinical practice.

Phase 3: Iterative Design and Prototyping

In phase 3, the research team presented to clinicians its conceptual sketches of ART and invited the clinicians to identify the most favorable designs of ART and its therapy surface, and also to rank their top requirements for ART by referencing the sketches. According to the clinicians, the top requirements for ART were a tilting surface, a therapy surface that maintains an optimal position during therapy, a table that is stable and locks during patient use, and a more stable and reliable raising-and-lowering mechanism. Not surprisingly, clinicians reported a preference for the therapy surface being used exclusively in a therapy room with a clinician present, and not at home by a patient unassisted, where home use of such assistive technology might encroach on clinical practice.

Phase 4: Formative Evaluation with Higher-Fidelity Prototypes

In Phase 4, the clinicians evaluated features of alternative ART prototypes and therapy surface prototypes, offering their ratings of *likes* (Is the user interaction with the prototype agreeable?), *needs* (What's expected of ART-patient interactions?), *usability* (How easy is ART to use?), and *anticipated problems* (What might go wrong in ART-patient interactions?). In their evaluations, the clinicians assumed the role of low-, medium-, and high-functioning stroke patients to capture the breadth of patient-ART interactions (table 7.1). Early in phase 4, the research team tested medium-fidelity cardboard prototypes, and later in phase 4, the team tested higher-fidelity prototypes. The research team learned the following critical points from phase 4: that low-functioning patients might require the assistance of clinicians or family members to maneuver the ART prototype; that the therapy surface was recommended for use only with clinician supervision; and that the therapy surface would require an arm restraint, should afford additional arm movements, and should record (by video camera or embedded sensors) clinical data of the rehabilitation process, such as degrees of limb movement, the ranges of motion, and forces applied by the patient. Overall, the research team learned that the clinicians were neutral as to whether they would use the therapy surface prototype in their clinical practice; curiously, the same clinicians reported that they imagined that the therapy surface would improve therapy sessions. As might be expected, while the clinicians recognized the promise of this assistive technology within the clinical space, they also recognized its impact on clinical practice and careers. (Something like ART might replace some of them.)

Phase 5: Summative Analysis
In phase 5, we invited eleven health-care providers, including physicians and occupational and physical therapists, to engage in "real world" interactions with our next iteration of a fully functioning ART prototype, where its various components were put to use within a double-occupancy patient room. As part of this study, clinicians were asked to conduct a therapy session on a (role-playing) patient's left upper extremity using our next iteration of the fully functioning therapy surface. The health-care providers evaluated individual components of ART, as in the previous phase, according to *likes*, *needs*, *usability*, and *anticipated problems*. As well, health-care providers were asked to describe their overall impression of ART in brief terms (i.e., a few words). Health-care providers subsequently repeated eleven terms more than once, with five participants offering that ART was "hard to use," and three offering that it was "useful." Overall, health-care providers reported frustration with the maneuverability of ART but offered a favorable review of ART's adjustable legs, its table controls (for raising and lowering the table surface), and the functionality of its therapy surface in supporting clinical sessions involving rehabilitation. Participants described ART as "easy to adapt to," "exciting," and "a new design ... that will take us into the next century," but ultimately "not quite there yet."

ROBOTICS THAT GIVE
Our five phases of investigation make evident that *home+* demands a most particular, intimate interaction between humans and cyber-physical systems; and ART's therapy surface, in particular, requires a delicate but exacting "touch" to impart confidence to patients and their caregivers that it can assistant them. To meet these demands, rigid-link robotics is not a likely candidate. The shortcomings of industrial, rigid-link robots for unstructured environments, where we live our everyday lives, have already been considered in chapter 4 in the context

of AWE. To summarize, the rigid-link robots commonplace in factories are composed of a series of heavy links, few in number, actuated by electric motors. The movements of rigid-link robots are correspondingly stiff but precise, which contrasts significantly with the less distracting, nuanced, and even graceful movements exhibited by AWE's display wall. Nevertheless, both the common industrial robot and AWE's display wall are based on a rigid-link, electromechanical system that is heavy and—worse for unstructured environments—unyielding upon contact with another physical mass. In simplest terms, these rigid-link robots don't "give" when they collide with people and their physical property. (Ouch!)

Consider, then, two of the most unstructured environments inhabited by two of the most vulnerable human populations, both of which define the *home+* context: dwellings inhabited by older people and health-care facilities occupied by patients. In these two demanding contexts, architectural robotic artifacts must not only move in a manner that is safe, comprehensible, and pleasing to their users but also be able to "give"—to be compliant upon contact with people and physical things. In such an intimate setting as *home+*, robotics will inevitably collide with people and their physical property; consequently, they must be compliant to avoid costly and potentially grave consequences.

Compliant, fluid, graceful kinematics is characteristic of the robotics classification called *continuum robotics* exemplified by ART's therapy surface. An emerging subfield, continuum robotics describes the kinds of robotics with smooth, compliant backbones that render their movement fluid, natural, and more life-like. Overall, the smooth movement and softness of continuum robots lends them to intimate and elicited interactions with users, while stiff and hard, rigid-linked robots are better suited to the demands for strength, repetition, and accuracy of industrial applications. As a flexible, continuous two-dimensional (2-D) programmable surface, ART's therapy surface represents a new class of continuum robotics capable of contributing to the formation of physical space within the built environment scale.

How does ART's therapy surface compare to AWE's display wall? As considered in chapter 4, AWE's display wall, having more degrees of freedom than it requires to move between configurations, is classified as a *redundant* robot; it has many ways "to go from here to there." But if we imagine the number of joints in the display wall (or, say, in a watchband) approaching infinity, we then have a *continuum* surface. This display wall of countless links is, in this way, "essentially invertebrate" in the way that a snake is essentially invertebrate, given its larger number of smaller links that bend at distinct and well-defined points. (See figure 7.9, in which AWE's display wall is extended by a number of additional panels to form a "snake-like" room enclosure.)

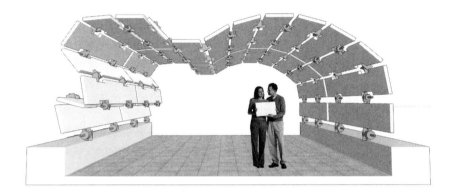

7.9
AWE with many more links becomes a
snake-like room.

The therapy surface of ART differs, however, from the AWE "display wall of countless links" in a fundamental way: while AWE's display wall bends only in one direction (i.e., it bends in section), ART's therapy surface bends in more than one direction (it morphs across its surface). Consequently, the therapy surface can assume a multitude of shapes that aren't stiff like commonplace industrial robots or limited (as is the AWE display wall) by bending in section alone. Rather, ART's therapy surface, and with it, the Interactive Wall of TU Delft's Hyperbody Research Group (figure 7.10), smoothly morph in two dimensions (like the sea). Given their continuous and highly maneuverable backbone, 2-D continuum robots like ART's therapy surface and the Interactive Wall are classified as *invertebrate*. The control of such invertebrate, continuum robots—the use of sensors, actuators, and algorithms to configure them—is a difficult matter. Unlike the control of rigid-link robots, the control of invertebrate, continuum robots often involves controlling for not only bending but also the extension and contraction of the physical mass, compared to simply controlling the hinging of an axial joint in the typical rigid-link robot composed of a limited number of joints.

7.10 (right)

Interactive Wall (Hyperbody Research Group, 2009), another 2-D compliant surface at the scale of the built environment (photograph by Walter Fogel; reprinted with permission of Festo AG & Co. and the Hyperbody Research Group, Technical University of Delft).

Let's consider the bending in two different continuum robots in concept: one of these is one-dimensional (1-D), and the other is 2-D. If you take two short segments of a continuum robot and connect them, head to head, you then have a continuum robot in the form of an extended line (much like an elephant's trunk). The bending that occurs within each segment of such a linear robot is defined as one-dimensional. (A well-known example of this kind of robotic appendage is the OctArm developed by Ian Walker, my close research collaborator.) To make the kinematics of a one-dimensional (trunk-like) continuum robot relatively less cumbersome to define, it *theoretically* reconfigures as the "essentially invertebrate" snake just described, having a very large number of smaller links that bend at distinct and well-defined points in such a way that each of its segments bends at constant curvature. If, instead of connecting segments head-to-head, you connected them in such a way that they form a surface (e.g., three segments to make a triangular surface, four segments to make a quadrilateral surface), and then co-join several such surfaces, you have the continuum robotic surface exemplified by the compliant therapy surface of ART. The bending that occurs within each surface of this continuum robotic surface is defined as two-dimensional. As such, the kinematics for such a continuum robotic surface is highly complex, as a continuum robot has, theoretically, an infinite number of joints across a surface, rather than along a line (as in OctArm) and, consequently, an infinite number of ways in two dimensions to assume its goal configuration. To make the kinematics of a two-dimensional continuum robot surface (like ART's) relatively less cumbersome to define,

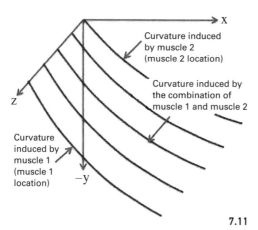

7.11
Diagram showing a compliant surface of constant curvatures.

it theoretically reconfigures as a surface of many snakes joined to one another, at their sides, so that an *area* of this surface has constant curvature (figure 7.11). Consequently, the problem of determining the shape of this robotic surface, its kinematics, is simplified compared to a determination made on the basis of the curvatures of a considerable number of very short segments of the surface. But even when characterizing the ART surface (and continuum robotics generally) as essentially invertebrate robots exhibiting constant curvature, the number and complexity of terms involved in their kinematics can be formidable. (The kinematics for such a surface is reported in a technical paper authored by our research team.[31])

A more tangible way of understanding continuum robotic surfaces, and so, ART's therapy surface, is through the qualitative study of functional, physical prototypes. To realize ART's therapy surface, my research team built a working prototype from a square surface of conventional foam (36 centimeters on a side), two McKibben actuators, some zip ties to fasten the actuators to the foam surfaces, and a Kinect from Microsoft's computer gaming system. McKibben actuators are artificial muscles (i.e., bladders) that expand and contract as they're filled or depleted of compressed air. According to the recipe advanced jointly by my lab and by the co-joined lab of my close collaborator, Ian Walker, we built our own McKibben actuators for AWE using high-temperature, silicone rubber tubing, expandable mesh sleeves, nylon reducing couplings, and nylon, single-barbed, tube-fitting plugs (figure 7.12). Digital pressure-regulators control the precise pressure of air delivered to the actuators. Flexible tubing connects the system of actuators, regulators, the air compressor (a common air tank will do), and the control computer (a common laptop will do). We attached two actuators to the square surface of sheet foam in four different orientations: running the actuators in parallel, forming a perpendicular ("Swiss") cross with them, and making 45° and 90° angles of them. We marked twenty-five points on the square surface that we tracked with the Kinect sensor, each point being either 8 or 10 centimeters from the previous point. For each muscle arrangement, two sets of data were collected. The first set of data measured the depth (the distance of the point on the surface from the sensor) of the un-actuated surface. The second set measured the depth of the points on the actuated surface by using a computer program that allowed the user to select feature points in two different image frames. The selected points in the first image frame captured the depth for the un-actuated surface at each point; and the selected points in the second image frame captured the depth for the actuated surface at each point. With respect to the surface, the Kinect sensor was positioned directly in front of it, at the same height of it, and parallel to its surface. To calculate the distance that each point on the surface moved in the x-direction (i.e., the height that the surface

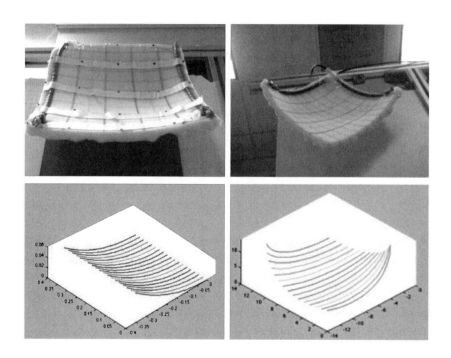

7.12 (above and opposite)
Evaluating the therapy surface's kinematic models: qualitative versus simulated results.

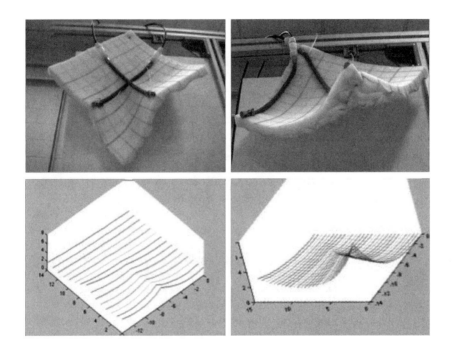

7.13
ART's therapy surface—working prototype.

7.14
Understanding the behaviors required of ART's therapy surface.

moved), the actuated data was subtracted from the un-actuated data. Once the x-distance data had been calculated for each point, the x-distances for the same points were calculated using the appropriate (i.e., muscle-arrangement dependent) kinematic model; then, the mean square error (MSE) between the two distances was calculated. Additionally, the MSE between the same kinematic data and the physical data for an un-actuated surface, which would have x-distances of zero, was calculated for comparison. This MSE would represent the worst-case scenario and should, therefore, be larger than the MSE for the actuated surface. If all this is a bit difficult to follow, figure 7.12 shows that the theoretical kinematics for the different muscle arrangements reasonably approximated the data we collected from testing the physical surface.

The analytic exploration just described guided the more complex design of ART's therapy surface. Rather than two actuators bending a square sheet of foam, the therapy surface (figure 7.13) was actuated by multiple actuators and required exacting behaviors that match five therapeutic exercises for the upper limbs performed by post-stroke patients and guided by physical therapists working without any medical devices during physical therapy sessions.

Before designing, prototyping, and testing such a therapy surface, there were starting points for us to consider. The first of these starting points is that an assistive robotic technology for rehabilitation must be recognized as "an advanced tool available to the physiotherapist, not a replacement" for the physiotherapist.[32] This point is "contrary to public perception," which envisions robots *replacing* human workers.[33] Contrary to this common misperception, the goal for us was to develop the therapy surface and ART overall as a tool having the capacity to "automate labor-intensive training paradigms" and even perform therapy movements that therapists may not be capable of performing.[34] With the therapy surface, we aspired to make the therapy experience more efficient and productive, and maybe more meaningful for patient and caregiver.

A second starting point for us was to create an assistive interface that mimics the behavior of a human therapist (figure 17.14). Like a human therapist guiding therapy practice, the therapy surface (like all rehabilitation robotics) needs to "be compliant when assisting movement, provide full support within the patient's passive range of motion, and nourish the patient's confidence and motivation levels through goal-oriented informative biofeedback."[35] Assistive robots like ART should mimic the human caregiver in these prescribed ways for two reasons: first, to make patients more comfortable with the device; and second, to make therapists more comfortable in allowing patients to use the device. Ultimately, such assistive robotic technology is intended to provide the therapist more time to focus on interpreting the patient's progress.[36]

A third starting point for us was the identification by robotics researcher Rui Loureiro and colleagues at UCL (University College London) of five main characteristics of upper-extremity rehabilitation robots that must be addressed in their design:

1. *The classification of the robot's therapy technique* Rehabilitation robots can be designed to perform *passive*, *active*, and/or *interactive* therapy. In passive therapy, the robot does not actually actuate the patient's joints; it simply maintains the limb within a specified range. In active therapy, the robot moves the person's limb through a predefined motion path. In interactive therapy, the robot reacts to the patient's movements. Each of these systems is beneficial in its own way, and each requires different hardware to accomplish the desired tasks. For ART's therapy surface, we decided to transition the therapy technique from active to a combination of active and therapist-interactive. Our intent was to have two different modes in which the device can act: in the first mode, the therapist can choose a predefined movement sequence and input the range of motion for the patient; and in the second mode, the therapist can adjust the pressure levels of the pneumatic muscles by turning ART's (linear potentiometer) knobs, adjusting the curvature of the surface and thereby adjusting the degree of wrist extension, wrist flexion, or arm cupping. This provides the therapist more possibilities to meet the needs of rehabilitating patients.

2. *The identification of joints to be rehabilitated* For upper-extremity rehabilitation, the therapeutic focus is on movements of the shoulder, elbow, forearm, wrist, and/or hand. An upper-extremity rehabilitation device like ART's therapy surface does not have to perform therapy on all of the aforementioned body parts. Through our studies with participating clinical staff, we designed for five therapeutic exercises: supination, pronation, wrist flexion, wrist extension, and arm cupping.

3. *The selection of robotic technology to be used* Rehabilitative robots may be classified by type as *unilateral*, *bilateral*, *partial-exoskeletal*, *full-exoskeletal*, *wire-based*, and *multirobotic*. These subcategories have been combined in multiple ways to help improve the safety and performance of robots in their usage over time. ART's therapy surface may be classified as a unilateral rehabilitative robot whereby the unaffected arm is suppressed from use, forcing the affected arm to enact the therapeutic movements. Additionally, ART's therapy surface was designed as a complement to the larger ART artifact (i.e., the table) that itself exemplifies a unilateral rehabilitative robot, guiding specific therapy movements by way of its own affordances (as elaborated later in this section). ART's therapy surface was designed, as well, as a detachable accessory to ART; in this way, the surface can operate independently of the table if the surface is connected to an accompanying mechanical column housing the hardware that activates it. This mechanical column would reasonably be the rear half of the

columnar body of ART (see figure 7.5) or the equivalent containing the same: an air tank, pressure regulators with air tubes, linear potentiometers, a single-pole, single-throw on/off switch, and a circuit board. Irrespective of the type and subtypes of the robot, a primary goal in designing the therapy surface was to ensin the clinical environment in which the robot will perform therapy.

4. *The physical location of the rehabilitation* Rehabilitation therapy may be designed exclusively for use in a clinical setting while a patient is in acute care or alternatively for home use. We evaluated ART's therapy surface in a patient's hospital room where the patient was sitting upright in his or her bed or wheelchair. We believe that the therapy surface, with minor adjustments, could also be used when the patient is lying down, and could be used in a home setting. Indeed, as integral to our vision of *home+*, we designed ART and its therapy surface for clinical and home use.

5. *The selection of the actuator type* There are numerous types of actuators that can be used in rehabilitation robots, divisible into two main categories: electric drives and pneumatic actuators. As considered already, pneumatic actuators were envisioned for ART's therapy surface for their nuanced, natural (human or life-like) movements, for their compliance or "give," and also for their high power-to-weight ratios.

In their paper, Loureiro and colleagues use the five characteristics just considered in their examination and evaluation of upper-extremity rehabilitation robots developed to date, while we used the same five characteristics in designing and evaluating our own upper-extremity rehabilitation robotic surface.

With these starting points in mind, the research team iteratively designed and evaluated the therapy surface in tandem with the larger five-phase ART investigation (as was elaborated here), translating the early studies (with the foam square and two actuators) into a rehabilitative robotic surface. Notably, in phase 1, specifically through observations of therapy sessions, the research team drew three inferences that would significantly affect the therapy surface's design requirements. First, we inferred that our continuum surface could perform many therapeutic exercises, including wrist flexion and extension, forearm pronation and supination, flexor synergy, shoulder flexion, shoulder rotation, cross-body movements, as well as arm cupping for support. Second, we inferred that these same exercises were sufficiently commonplace in therapy sessions so that our therapy surface would prove useful. Third, we inferred that our surface could provide enough variability in the movements of these exercises to accommodate various injury levels and types.

In our design development, we ultimately divided the therapy movements into those that can be achieved by our continuum robotic surface (wrist extension and flexion, forearm cupping, forearm pronation and supination; see figure 7.15) and those that can be achieved simply by the movement of ART on which the continuum surface was mounted (flexor synergy, shoulder flexion, shoulder rotation, and cross-body movements). To assess how well the various behaviors of the robots matched the expectations of the therapists for each of these exercises, we prepared and presented a video of the various therapy movements performed by the prototyped surface to a group of therapists and asked them the following:

- Is this how you would perform wrist flexion?
- Is this how you would perform wrist extension?
- Does this device offer enough variability for various patient types?
- Would you use this device?
- Do you think a patient would use this device?
- Do you think this device is safe?
- How does this device improve therapy sessions?
- What information would you want this device to gather?
- What other therapy movements do you think the device can perform?

The five iterative phases of design research resulted in a comprehensive prototype of a continuum robotic surface for stroke rehabilitation therapy that gathered positive and promising responses from the clinicians. Indeed, we designed ART and its therapy surface not specifically for clinical or home use but rather as integral (and key) to our *home+* vision in which house and clinic are essentially replaced by a physical environment for well-being.

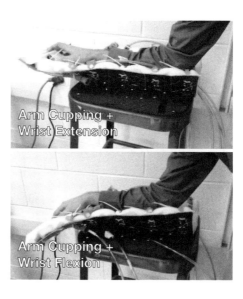

7.15
ART's therapy surface performing therapies.

THINKING ABOUT ARCHITECTURE AND ENGINEERING DIFFERENTLY

As just considered for ART's therapy surface, soft, smooth, fluid reconfigurations are essential to the success of architectural robotics as it supports and enhances everyday human activities in, very often, intimate confines. Figure 7.16 provides a summary of four "soft robotic" efforts considered in this book, conceived by me and my close collaborator, Ian Walker, and which are counter to a rigid-link robotics approach. Similarly, figure 7.17 captures in graphic, even macabre terms our concept of the distributed environment typology of architectural robotics, where we imagine a humanoid robot being dismembered so that its limbs, head, and torso become a *living room* full of lively furnishings, each one tuned to specific tasks and playing its part, each one knowing something about the others and something about you.

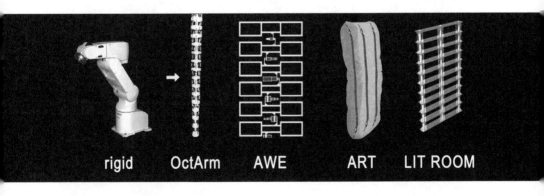

7.16

A summary of four "soft robotic" efforts considered in this book, which are counter to a rigid-link robotics approach (far left) (diagram by the author).

The ART of *home+*

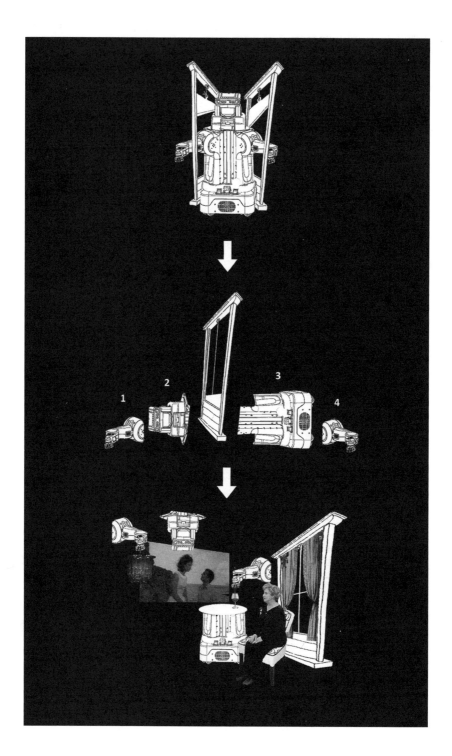

By now it should be evident that robotics embedded in the built environment opens up a very different way of conceptualizing architecture. The notion that the physical mass of an architectural work is set in motion by robotics, reconfiguring its spatial quality, disrupts our fundamental understanding of architecture as *firmitas*. The first of three defining attributes of an architectural work offered by Vitruvius, *firmitas* implies firmness, constancy, solidity, and stability. When the robotics embedded in the built environment is classified as redundant or continuum rather than rigid link, as in AWE or ART, the sense of *firmitas* is arguably more upended, as the architectural work is not only moving but, in a way, quivering as if it were alive, and shape-shifting into something that seems foreign to what it was at an earlier point in time.

Before ending this consideration of ART's therapy surface, it is useful to ponder the questions suggested by it, and the larger ART, and our vision of the *home+* to which they belong: What kind of place is this? How does it reflect our current state (or, you might say, disposition)? In a curious way, French philosopher Gilles Deleuze and psychoanalyst Félix Guattari together ponder the second question literally and abstractly in *Mille plateau* ("A Thousand Plateaus"). In the twelfth "plateau," exploring the relationship between the state and its military, Deleuze and Guattari posit two models of thinking and acting that, strangely, provide a compelling perception that applies well to architectural robotics, the kinds of places it makes, and how they reflect, interact with, and somewhat become us.[37] These two models are the *authoritative model* exemplified by the state, and the *vortical model* exemplified by the military. Architectural robotics, as offered here, is very much the latter, and architecture is very much the former.

The authoritative model is characterized by the straight line, the parallel (laminar) flow, solid things, and a spatial environment that is closed, striated, and sedentary. In its propensity to edify and monumentalize, architecture (and with it, engineering), argue Deleuze and Guattari, have "an apparent affinity for the State" and its authoritative model. The authoritative model is manifested in the built environment as "walls, enclosures and roads between enclosures," "bridges," "monuments," and, generally, a "metric plane" that is *"lineal, territorial, and numerical."*[38]

7.17
From a humanoid robot to a *living room*
(diagram by the author).

In contrast, the vortical model is characterized by the curvilinear line, the spiral (vortical) flow, the flow of matter, and a spatial environment characterized as open, smooth, flowing, and fluid.[39] Free flowing and measureless, the vortical model is manifested in the natural environment in the desert and the sea.[40] Outside of its manifestations in nature, the vortical model is exemplified by the intermezzo of music, the space of Zen, and in the built environment (in a rare exception) as the impossibly slender, sky-reaching Gothic cathedral.[41]

More generally, Deleuze and Guattari characterize the vortical space as "vectors of deterritorialization in perpetual motion."[42] Additionally, Deleuze and Guattari, referencing phenomenologist Edmund Husserl, describe the vortical space as a "fuzzy aggregate," "a kind of intermediary," "*vague*" and "yet rigorous."[43] Something of the same is central to the thinking of Manuel Castells in his concept of the "space of flows."[44] For Castells as for Deleuze and Guattari, the fluid, vortical "space" is a way of thinking and making that finds validation in our propensity for "neurological play," in the lure of controlled "ambiguity" that "engages and challenges the brain to allow multiple meanings."[45]

If this philosophical equation between form and thinking seems itself vague and ambiguous, we need only walk the linear boulevards that cut through Paris, Rome, and Washington, D.C., to connect urban form to figureheads (Napoleon III, Mussolini, George Washington). Today, we have not the grand boulevard but rather the kingly quarters of the private domain, exemplified by Apple's new Cupertino, California, headquarters. Accommodating 12,000 Apple employees, this $5 billion building designed by Foster + Partners (with Arup as engineers) takes the form of a perfect glass circle defining a perfectly circular courtyard. No matter how elegant its glass skin, this closed, *authoritative* form—an apple without the bite—chillingly evokes Dave Eggers's aptly titled *The Circle* (2013), which chronicles a tech employee caught in the web of a colossal information technology company.

The other side of this equation, the vortical, surfaces in the architectural writings and works of Robert Venturi (oftentimes in collaboration with his wife, Denise Scott Brown) in his theory of a "*both-and*" architecture rather than an "*either-or*" architecture: an architectural work characterized as *both* complex *and* contradictory, *both* high-culture *and* low-culture, *both* contemporary *and* drawn from history.[46] Much like Venturi but from the perspective of robotics and not architecture, Rodney Brooks, referenced earlier, envisions a future in the *both-and*—both organic and mechanical.[47] Suggesting that the convergence of human intelligence and machine intelligence will come only by way of a new conceptual framework, outside that of digital computing, Brooks calls for a biotechnical body, and sees its emergence in our becoming increasingly more robotic (by way of, for instance, advanced prostheses) while robots are becoming more biological (in their capacity to sense, adapt, learn, move, and otherwise actuate).[48] Presumably for Brooks, we will meet one day in the middle: human-machine, machine-human.

But unlike Venturi's dialectical, physically static, and aestheticized *both-and* architecture, and taking a more tempered view of Brooks's prophetic, biotechnical body, architectural robotics suggests a very different way of thinking about architecture and robotics today, whereby architecture is more than an aesthetic search or a stylistic path, and robotics is more than a technological quest. Something at the threshold between architecture and robotics and their long-standing concerns is captured in Deleuze and Guattari's "Concrete Rules and Abstract Machines," in which they define a novel, "technological plane" as "not simply made of formed substances, aluminum, plastic, electric wire, etc., nor of organizing forms, program, prototypes, etc., but of a totality (ensemble) of unformed matters which present no more than degrees of intensity ... and diagrammatic functions which only present differential equations."[49]

An early design exemplar of the distributed environment typology of architectural robotics, the *home+* vision, with ART as its key component, represents a start to realizing the dream of a *living room*, a distributed suite of robotic furnishings that "come to life," that know about us and know about each other. Furnished with this information about itself and its surroundings, the living room reconfigures, smoothly and softly, to support us in our everyday lives. But more than this, the convergence of architecture and robotics, manifested as a living room, promises new vocabularies of design and new, complex realms of understanding ourselves in this dynamic, expanding ecosystem of bits, bytes, and biology.

8   THE DISTRIBUTED ENVIRONMENT

The distributed environment (figure 8.1) that, seemingly, *comes to life* is composed of individual, furniture-like components forming a suite of physically distributed furnishings. Typologically, the components of the distributed environment may look like familiar furnishings (say, a chair, desk, or sofa) or alternatively may look like two or more familiar furnishings joined together or hybridized as a single unit. In any case, the single unit of furniture physically reconfigures by the means of embedded robotics in response to human interactions with it and/or by some input from the Internet or from its physical surroundings. Embedded robotics expands the affordances of such interactive furniture compared with the familiar affordances of static, single-purpose, conventional furniture.

As the contents of the distributed environment are identifiable as discrete physical units, you can very much recognize the room—the physical envelope or container—for what it is: most commonly, rectangular in plan and of a familiar height. Figure 8.1 is meant to capture the essential character of the distributed environment typology, in which the three cells in the figure, furnished with physically distributed, morphing furniture, represent three everyday instances during which inhabitants eat and speak, lounge, and play. (Compare figure 8.1 and figure 5.1 to see how architectural robotics accommodates these same human activities differently.) While the individual furnishings that constitute the distributed environment are physically discrete, each one is networked with the others and consequently knows something about the larger suite of furnishings, acting accordingly. In this respect, all three architectural robotics typologies considered in this book can be described as "distributed environments" in the sense that *computing* is distributed across them, but only the distributed environment is characterized by discrete *physical* components distributed spatially. One notable exception is when the furnishings of the distributed environment are characterized by modular robotics, in which case (some or all) of the furnishing units are capable of physically connecting to one another. Yet even in the case of modular robotic furnishings, the individual physical units (the physical modules) tend to be visually discernible to users as discrete entities.

An exemplar of the distributed environment typology, *home+* is defined by a collection of networked, cyber-physical devices, each having different functionalities tuned for different purposes to support a common human need or to exploit an opportunity to improve the lives of its inhabitant(s). *home+* is composed, on the surface, of the commonplace furniture of an ordinary, contemporary house. The transformation to *home+* is achieved, borrowing the words of William Mitchell, by way of "geographically distributed assemblages of diverse, highly specialized, intercommunicating artifacts" that render the

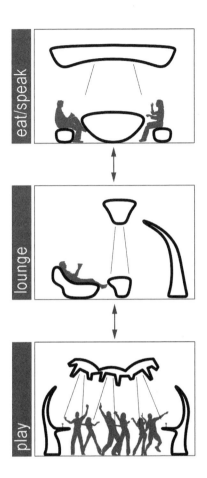

8.1
Diagram of the distributed environment typology (diagram by the author).

physical environment a "a robot for living in."[1] In similar terms, former *Wired* editor Kevin Kelly imagines a future artificial "ecology" of intelligent "rooms stuffed with co-evolutionary furniture" and a "mob of tiny smart objects," all having an "awareness of each other, of themselves, and of me."[2] Composed of multiple, commonplace domestic artifacts, the distributed environment manifests something of the behavior of swarms and also subscribes to the notion of the artificial life community that "in living systems, the whole is more than the sum of its parts."[3] *The remarkable feature of the distributed environment typology is that, seemingly, the familiar furnishings found at home, at work, at school, and within other everyday places "come to life," forming a "living" room.*

The Distributed Environment

III TRANSFIGURABLE

9 PORTALS TO ELSEWHERE

When we live in a manor house we dream of a cottage, and when we live in a cottage we dream of a palace. Better still, we all have our cottage moments and our palace moments.

—Gaston Bachelard, *The Poetics of Space*

The French philosopher Gaston Bachelard eloquently captured that restlessness, that *ennui* that so many of us experience. For those of us fortunate enough to have our basic needs met, we often find ourselves in one place, wanting to be somewhere else. This is the cliché of the country girl, dreaming of life in the big city. Or the urban sophisticate, dreaming of the ocean from midtown. Or the child—prisoner to his routine—fantasizing about an escape on the high seas to faraway lands.

Max is one of these. In the familiar children's classic *Where the Wild Things Are* (1963), young Max is sent to his room without supper as punishment for his childish misbehavior. It was "that very night, in Max's room, [that] a forest grew and grew—and grew until his ceiling hung with vines and the walls became the world all around."[1] Out of Max's imagination, a mysterious, wild forest and sea grew in Max's bedroom—the surreal, transformative landscape of his adventure. In *The Salamander Room* by Anne Mazer, young Brian—one step more willful than Max—actively transforms his bedroom into a habitat that can support the salamander he's captured from the wild. Before long, Brian's mundane bedroom becomes populated not only by his salamander but also by trees, a pool of water, a variety of birds, caterpillars, frogs, and other living things (figure 9.1).

Of course, the bedrooms of Brian and Max are the things of fairy tales. No matter how lively your imagination, no matter how restless you may be, you cannot *will* a static, conventional room to transport you, within its confines, to somewhere else. Or can you? With the advances of digital technology, something akin to the bedroom portals of Brian and Max has been achieved, reduced to a two-dimensional plane, within the experimental Microsoft Home developed in Redmond, Washington. Lamentably, the Microsoft Home, at least the one I toured several years ago, was not the transformative environment I had hoped to discover, but instead (as I should have expected) a collection of near market-ready technologies, like the interactive wallpaper that transforms the walls of, say, a teenager's bedroom into a wraparound display (figure 9.2). Here in this bedroom, we have nothing of the fleshiness of Max's forest or the torrid currents of his sea, but rather some changeable yet arguably lifeless images stretched over a large two-dimensional plane, providing us more the experience of conventional wallpaper than an imagined, other world. As I watched the digital

9.1

Like Max's bedroom, the Salamander Room is a boy's portal from his mundane surroundings to another world. (Detail of illustration by Steve Johnson and Lou Fancher, copyright © 1991 by Steve Johnson and Lou Fancher; from *The Salaman-der Room* by Anne Mazer. Used by permission of Alfred A. Knopf, an imprint of Random House Children's Books, a division of Penguin Random House LLC. All rights reserved.)

9.2

The digital wallpaper in the Microsoft House: *Is this it?* (Source:www.engadget.com/2011/05/03/microsofts-home-of-the-future-lulls-teens-to-sleep-with-tweets/.)

wallpaper display images of the bedtime classic *Goodnight Moon*, I was thinking how much I preferred, at least for my own twin children back home, lending my own voice, imaginings, and associations to the words of the printed book.

Keep in mind that, in Max's case, it is *his imagination* that brings the wondrous transformation to the room, not the software of an intelligent house (e.g., the user-selected e-book or YouTube video of the children's book or the video output of a recommender system). But what if Max was not only mentally engaged but also more physically engaged in transforming his room; that is, what if Max were more like Brian? Moreover, what if the transformations in the room were the outcome of an interaction between the room and some combination of its inhabitant's imagination, the inhabitant's capacity to physically alter the environment, and the room's sense of the inhabitant's current state?

Enter Duc Jean des Esseintes, "a frail young man of thirty, anemic and nervous, with hollow cheeks, ... and long, slender hands."[2] Long before the Microsoft Home and its digital wallpaper, this protagonist of J. K. Huysmans's fantastical novel, *A rebours*, was consumed by boredom and consumed, as a recourse, with physically reconfiguring his home in different ways to reflect his own state of being. Like Max's adventures, the escapades of des Esseintes are confined to the interior of his home, furnished and decorated more like a *boudoir* (see figure 6.1) than a young boy's room, as befits the duke. The more important difference between des Esseintes and Max is that des Esseintes and the interior of his surreal home have an intense reciprocity (if your imagination will allow): they are mutually and very busily seeking that balance of environmental and mental conditions that might bring them into sync—that might make des Esseintes feel *at home* with his most familiar physical setting *and himself*. And we must remember from Bachelard that des Esseintes and his environs will never sync (come to rest, come to equilibrium, come to balance) for very long. Indeed, des Esseintes, like most of the rest of us, certainly has his "cottage moments and his palace moments."

For instance, a bedroom for des Esseintes was "a place for pleasure, contrived to excite the passions for nightly adventure; or else ... a retreat dedicated to sleep and solitude, a home of quiet thoughts, a kind of oratory."[3] To arrive at either or both of these preferred psycho-environmental states, des Esseintes reworked the many components of his lavish bedroom, initially, as his sought-after retreat for sleep and solitude:

> He must contrive a bed-chamber to resemble a monk's cell in a Religious House; but here came difficulty upon difficulty, for he refused absolutely to endure his personal occupation the austere ugliness that marks such refuges for penitence and prayer. [...] He arrived at the conclusion that the result to

be aimed at amounted to this—to arrange by means of objects cheerful in themselves a melancholy whole, or rather, while preserving its character of plain ugliness, to impress on the general effect of the room thus treated a kind of elegance and distinction; to reverse, in fact, the optical delusion of the stage, where cheap tinsel plays the part of expensive and sumptuous robes, to gain indeed precisely the opposite effect using costly and magnificent materials so as to give the impression of common rags; in a word, to fit up a Trappist's [sic] cell that should have the look of the genuine article, and yet of course be nothing of the sort.[4]

Clearly, des Esseintes will find no resting place here. And he seems no more satisfied as he seeks the right ambiance, in the same bedroom, for entertaining at night his female guest. He strains

... to envelope [her] in a vicious atmosphere, shaping its furniture on the model of her charms, copying the contortions of her ardour, imitating the spasms of her amorousness in the waving lines and intricate convolutions of wood and copper, adding a spice to the sugar-sweet languor of the blonde by the vivid, bright tone of its ornamentation, mitigating the salty savour of the brunette by tapestries of subdued, almost insipid hues.[5]

Unlike Max, des Esseintes can find no way out of his predicament, which has effectively imprisoned him in his bedroom. As told by Huysmans over a great many pages in *A rebours*, the duke attempts all kinds of elaborate reworkings of his environment, and the environment overall yields to him, but nothing it offers, nothing he organizes within it, feels just right to him. We find the same restless activity occurring in André Breton's *L'amour fou*, in which the poet enacts a frenetic yet surely futile ritual, aimed at capturing an absent woman's attention, that begins with his "opening a door, then shutting it," and proceeds with slipping "the blade of a knife randomly between the pages of a book" and "arranging [objects] in unusual ways."[6] Like Breton, surely, we are not expecting that this series of domestic reconfigurations will result in her arrival. But the point here is that the room can serve as a vehicle to reflect the dynamic and often restless mental states of their inhabitants.

Both of these tales have roots in the classical Greek myth of Pygmalion, but not as Ovid had it, in the relatively quick and easy marriage between sculptor and his sculpture, where the bride is the sculpted figure of a woman that has miraculously come to life before him. Instead, the tales of Huysmans and Breton are rooted more in the medieval retelling of the Pygmalion tale by Jean de Meun. In Jean de Meun's version, we witness the sculptor's ceaseless dressing and redressing of the sculpted female form. Nothing suits her; nothing suits him.

The sculptor's work and his emotional state find "no peace or truce" as if the sculptor's very emotional state were made of clay.[7]

From the canon of architecture, a "Pygmalion" not unlike that of de Meun appears in a most peculiar passage of Filarete's *Treatise on Architecture*. Here, at the near conclusion of this Renaissance treatise, Filarete considers best practices for developing hand drawing, advocating that the architect arrange the model as follows:

> When you have to clothe a person, do as I tell you. Have a little wooden figure with jointed arms, legs, and neck. Then make a dress of linen in whatever fashion you choose, as if it were alive. Put it on him in the action that you wish and fix it up. If these drapes do not hang as you wish, take melted glue and bathe the figure well. Then fix the folds as you want them and let them dry so they will firm. If you then wish to arrange them in another way, put it in warm water and you can change them into another form. Draw your figures in the way you want them to be dressed.[8]

The way to cultivate an architect, Filarete advocates, is not to rest on our hunches but to tirelessly reconfigure the external, physical state until it reflects our own wants, needs, and desires. A more contemporaneous version of this Pygmalion-like tale for architecture comes from the Turinese architect we met in chapter 3, Carlo Mollino:

> Every aesthetic activity is therefore technically reduced to this activity of selection, transfiguration, and coincidence with our sensations. This activity is affirmed in the joke: "The block of marble is the most beautiful of all statues." What is said here of marble and statues—that within the material we can conceive its infinite potentials—can be extended further to all of the conditions of art: that in light, there are all colors; in space and volumes, all of architecture; in words and in language, all poems.[9]

Mollino recognized innumerable prospects within architecture.

The lesson of such stories, whether Max's flight of fancy or des Esseintes' travails, is their sharp focus on that intimate relationship between the inhabitant and the room, the state of the environment and the mental state; and, in *A rebours*, *L'amour fou*, and the retellings of the Pygmalion myth, how dynamic is the interaction between the inhabitant and the room, a tireless dance of the animate and the inanimate; and finally, for Carlo Mollino, the discovery of all of architecture within a space or volume.

How often are we restless in environments familiar to us (our homes, workplaces, our hometowns and cities) and yearn to be elsewhere (familiar or foreign).

Above all, what we learn from the tales considered here is that we can actively plot our escape from these oppressive interior environments without leaving their confines. If we long for a place far away to feel more at home, perhaps we can make our home feel more like a faraway place.

In "The Ethical Significance of Environmental Beauty," Karsten Harries recognizes the same complicated yearnings for home in both the occasion of the first Moon landing and the release of Stanley Kubrick's *2001: A Space Odyssey* one year earlier. Captured in Kubrick's film, the complicated feelings accompanying the first Moon landing arise from

> ... two seemingly contradictory desires: on the one hand, by a desire for the sublime, by a longing for the excitement of encountering something totally other than our all too familiar, often so confining world; something numinous, a sublime *mysterium tremedum et fascinans* that would answer to a freedom that refused to be bound to the earth; on the other hand, by a desire for the beautiful, by a longing to encounter, even out there, an intelligence much like our own so that, instead of feeling lost in space, we could once more feel at home in this new so greatly enlarged cosmos.[10]

A half-century after the first Moon landing and the release of *2001: A Space Odyssey*, we have grown accustomed (or *habituated*, more aptly) to spaceflight and the elaborate special effects of cinema. And maybe some of us have grown weary of *the journey*, of reaching so far. (Indeed, the last landing of humans on the Moon was on December 11, 1972, by *Apollo 17*.) Lamentably, today's technologies and the powers that wield them have not since landed a human on Mars (or returned one to the Moon), but immerse us instead in the otherworldly experience of digital media, permeated with the frightfully familiar trappings of consumerism. Is this the destination? Is this it? As a portal to elsewhere grounded in this world, architectural robotics is no Mars nor spacecraft but a built environment promising to be a sensuous, dynamic, capacious, expansive, "omnisensorial" abode, rooted in yet cantilevered out far from the information world.

WE MAKE THE SPACE

A *portal to elsewhere* is a reconfigurable room that is not fragmentary or empty but filled with human activity. In other words: in this small room, a lot happens. The Italian word for room, *stanza*, implies a place *to stay*, a capacious vessel that also is the name for a grouping of lines in poetry, the space where lines of poetry are collected and displayed. If this portal is in fact a room filled with activity (and even poetry sometimes), then what happens outside it? More broadly, how does one draw the boundary between the space of architectural robotics and the space that isn't? The answer lies in what makes a space *a place*. And what makes a space *a place*—something meaningful to us—is the human acts that occur within it.

But there remains the question of what happens just on the other side of this place held together by human activity. Again, philosopher Henri Lefebvre helps us look at the other side. As Lefebvre offers, "Walls, enclosures and façades serve to define both a scene (where something takes place) and an obscene area to which everything that cannot or may not happen on the scene is relegated: whatever is inadmissible, be it malefic or forbidden, thus has its own hidden space on the near or the far side of a frontier."[11] As a consequence of the two-sided space, "confusion gradually arises between space and surface, with the latter determining a spatial abstraction which it endows with a half-imaginary, half-real physical existence."[12] For Lefebvre, space is fundamentally not what it seems, and surface is not altogether real.

As manifested in an architectural robotic environment, what *portals to elsewhere* can do, then, is to make partly visible the incoherency of space in a controlled manner that strives to prove delightful or at least stimulating to its inhabitants. In this kind of transfigurable environment, inhabitants are indeed a little less certain where they are compared to their typical, routine experience of their everyday places, but it is by way of the inhabitants' inhabiting such curious places that the inhabitants maintain control over them. Or, as Lefebvre maintains of such spaces, "only an act can hold—and hold together—such fragments in a homogenous totality. Only action can prevent dispersion, like a fist clenched around sand."[13]

Really, we only imagine the spaces we inhabit every day to be coherent, stable, and logical (as in Cartesian space); outside the working of human reasoning, space is, in Lefebvre's words, "incoherence under the banner of coherence, a cohesion grounded in scission and disjointedness, fluctuation and the ephemeral masquerading as stability, conflictual [sic] relationships embedded within an appearance of logic."[14] Accordingly, a physical, spatial environment is not a fact, but a construct: the production of its inhabitants as "conceived, perceived, and directly lived" by them.[15]

Ultimately, space is what we make of it. What architectural robotic environments can do is to invite its inhabitants to be transported through active engagement in its spaces. And, as elaborated in the chapter that follows, this immersion in spatial complexity is put to play, quite literally.

10 A LIT ROOM FOR THE PERSISTENCE OF PRINT

Ceci tuera cela ("This will kill that"). These are the famous words Victor Hugo used in *Notre-Dame de Paris* ("The Hunchback of Notre-Dame") to declare the triumph of printed media over architecture. With the growing use of the printing press in the Enlightenment, the printed book, and no longer the building, would become the primary communicative means of society. Needless to say, architecture was never beheaded at the *Place de la Concorde*, and its communicative power and currency has not since been entirely lost. Evidence to the contrary is indeed demonstrated today by the enduring phenomenon of "starchitecture," architectural works designed by a select group of celebrity architects determined to lend to their sponsors highly visible, high-culture recognition. Of course, starchitecture is not a phenomenon altogether distinct from Bernini serving Pope Urban VIII or Hardouin-Mansart serving Louis XIV. Irrespective of the leader's sphere of influence—nationhood, commerce, or religion—architecture has endured as a weighty communication technology, despite the ubiquity of the printing press.

And now, the Internet. As memory is short, there are many voices today warning us that the Internet will supplant older communication technologies—namely, printed media (periodicals in particular), television, and radio—just as the printed page was projected to usurp architecture. Surely, periodicals, television, and radio are being recalibrated for the digital age. But while computer displays proliferate in number and size (smaller and larger), printers are busy printing, the radio's playing, and televisions are turned "on" far too long in a day for anyone's own good. Similarly, in higher education, despite the growing number of massive open online courses (MOOCs), universities are not lacking for students, and maybe the more promising future for higher education is not *either-or* but the hybridization of traditional and digital delivery systems. And while many of us, connected to the World Wide Web 24/7, no longer need to leave home for work—no longer need to get out of bed, really—some of us biked to work today, some of us went to the gym, and all of us touched things, manipulated them, and, in so doing, came to understand a great deal about the world we live in and who we are within it.

For many human-centered technological applications today, hybrid analog-and-digital systems are the likely victors over any singular communication medium or technology, analog or digital. Concurring with this proposition, Hiroshi Ishii, for one, laments that "HCI design" suffers from a "lack of diversity of input/output media, as well as too much bias towards graphical output at the expense of input from the real world."[1] A recourse to this "lack" in human-computer interaction (HCI) comes in the form of "embodied interaction" offered by Paul Dourish; and, later, Alissa Antle's concept of "embodied child-computer interaction."[2] For Antle, "meaning is created through restructuring the spatial configuration of elements in the environment."[3] Architectural robotics, particularly for learning applications, takes Antle's words quite literally, envisioning a reconfigurable, cyber-physical environment as coadaptive, allowing environments and their inhabitants to change and develop mutually. As Antle offers for specifically children—the target population soon considered in this chapter— "an environment ... that supports multiple spatial configurations promises to advance a child's grasp of our universe through active, creative exploration."[4] The arguments of Dourish and Antle are themselves supported by the child-initiated *Reggio* method of learning acquisition that predates their publications. The *Reggio Emilia* philosophy reminds us that "the environment is a teacher": that the physical environment in which children learn is central to fostering their capacities.[5]

"The lesson," stresses Michio Kaku in the *Physics of the Future*, "is that one medium never annihilates a previous one but coexists with it."[6] According to Kaku, the reason for this mix of media, old and new together, is the reliance of "our ancient ancestors," for their very survival, on sure-footed, physical evidence, not hearsay. Kaku argues that "in the competition between High Tech and High Touch ... we prefer to have both, but if given a choice we will chose High Touch, like our caveman ancestors."[7] Comparing our caveman ancestors with human beings today, writes Kaku, "we see no evidence that our brains and personalities have changed much since then."[8] Consequently, social networking sites are popular today as gossip was central to the lives of our earliest ancestors. This case of "ancestral gossip to social media" epitomizes Kaku's "cave man principle," which states "that if you want to predict the social interactions of humans in the future, simply imagine our social interactions 100,000 years ago and multiply by a billion."[9] A large part of this chapter, and this book, is dedicated to retaining old and new technologies and modes of human-machine interaction, to having these coexist, and to making room in the design of an increasingly cyber-physical environment for new technologies and new human-machine interactions anticipated and not yet imagined.

BACK TO THE BOOKS OF CHILDHOOD

Despite our digital addictions (or our digital drunkenness?), it is remarkable that the practice of reading aloud to children from a printed picture book, a book of images and words, remains a cornerstone of child-rearing practice. Parents across our most digitally literate nations fondly recall reading, and having read to them, one of the earliest picture books, Beatrix Potter's *The Tale of Peter Rabbit*, along with favorites like Dr. Seuss' *The Cat in the Hat* and Maurice

Sendak's *Where the Wild Things Are*. And these digitally literate parents, today, are still reading these same printed books to their own children as a means of cultivating literacy.

Meanwhile, the ascent of the personal computer, the smart phone, and the Kindle have not yet been accompanied by an equivalent rise in literacy rates anywhere. Illiteracy is not a third-world problem but a global problem. UNESCO estimates that roughly 20 percent of the world's population is illiterate.[10] And despite its wealth and its high ownership rate of digital smart devices, the United States has an alarming illiteracy problem: 14 percent of the nation's residents are illiterate, 32 million residents cannot read the directions on a medicine bottle, and 50 million residents cannot read above the fifth-grade level.[11] Meanwhile, within and outside the United States, Internet access and digital literacy—how people find, use, summarize, evaluate, create, and communicate information while using a broad range of digital devices—are also distressing problems. Clearly, ownership (of digital devices) and know-how are not equivalents. In whichever form, reading or digital, illiteracy affects both economic growth and the quality of life, locally, nationally, and globally.

With urgency, illiteracy is a challenge apt for the research community of information and communications technology (ICT), to which architectural robotics, HCI, and, broadly, computing belong. If too many children are spending too little time with printed books and too much of their time facing television and computer screens, perhaps there is a way to bridge the worlds of printed and digital technologies as a *portal* back to the books of childhood—back to that cornerstone of literacy, the *read-aloud*, in which adults read to young children, transmitting the love of reading along with the imagined worlds that reading conjures.

ABOARD THE CYBER-PHYSICAL

As a means for transporting us to the world of books and to the larger realm of the imagination, today's digital technologies, robotics in particular, offer still more than Microsoft's digital wallpaper, or the contrivances of the three French protagonists—des Esseintes, Breton, de Meun—of the previous chapter, or, closer to architecture, the Romans, who, "in the [architectural] orders ..., saw only a decoration which might be removed, omitted, displaced, or replaced by something else."[12] Today, the "simulated environments" envisioned by so many cultural fields for so long, and initially conceived in the research domain of intelligent environments by Nicholas Negroponte and the Soft Architecture Machine Group, are more accessible than ever.

Inspired by Negroponte's vision of a "living room that can simulate beaches and the mountains,"[13] the LIT ROOM exemplifies the transfigurable environment

10.1
The LIT ROOM, engaged by children and a librarian reader (photograph by the author).

10.2
The LIT ROOM (front-entry view), inside the Children's Reading Room of the Richland Library in Columbia, South Carolina (photograph by the author).

typology of architectural robotics, a "portal to elsewhere" embedded in the physical space of the library. As a robotic room, the LIT ROOM (figure 10.1) is transformed by words read from a picture book so that the everyday space of the library "merges" with the imaginary space of the book. *The book is a room; the room is a book.* A mixed-technology system for enhancing picture-book reading, the LIT ROOM combines the printed page with a multimodal, programmable experience evoking the book being read. The room-filled audio-visual-spatial effects of the LIT ROOM contextualize language and provide feedback to the participants. The LIT ROOM aims to scaffold critical literacy skills such as vocabulary acquisition, reading comprehension, and print motivation by creating a fun, interactive experience for children.

In the plainest terms, the LIT ROOM is a suite of four panels each having the dimension of a very wide door or a very tall window (3½ feet wide by 1 foot deep by 7½ feet high), supported by customized Bosch Rexroth aluminum framing that forms a room-scaled, rectangular volume, 12 feet wide by 12 feet deep by 8 feet high (figure 10.2). Embedded in each of the four panels are shape-changing, continuum-robot surfaces using tendons for their reconfigurability. Additionally, the four panels are embedded with LED lighting, audio speakers, and associated electronics. At the center of the installation is a small, low-lying, circular table housing a conventional tablet computer. This table with embedded tablet operates as a lazy Susan, rotating about its center axis to provide access to it for all LIT ROOM participants gathered around it. All of what composes the LIT ROOM sits within a conventional library interior (figure 10.3) or within some other appropriate, physical environment.

More conceptually, the LIT ROOM is inspired partly by the arguments offered earlier from Ishii, Antle, and *Reggio Emilia*. Following from Ishii's sentiment that HCI design suffers from a "lack of diversity of input/output media" and "a bias towards graphical output at the expense of input from the real world," the LIT ROOM conflates physical space and cyberspace while providing a rich media palette for play. Following from Antle's concept of creating meaning "through restructuring the spatial configuration of elements in the environment," the robotics-embedded LIT ROOM environment is physically reconfigured by children (figure 10.4), rendering it coadaptive: in course, both the environment and its inhabitants are transformed. This coadaption in the LIT ROOM effectively reworks the *Reggio* philosophy, making it reflexive: so while "the environment is a teacher" to the child, the child is given ample opportunity to teach or cultivate the form the environment.

In conceiving the LIT ROOM, my research team and I hypothesized that a transfigurable environment, one that transports you elsewhere, cultivates learning. For this, we found validation, on a fundamental level again, in the findings

10.3

The LIT ROOM in the library, side and top views (photographs by the author).

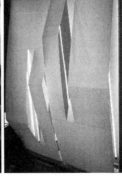

10.4

(Left) Children reconfiguring the LIT ROOM using our current tablet interface to evoke the book at room scale—in sound, color, form, and movement. **(Right)** Detail of one of the four robotic panels (photographs by the author).

previously considered in chapter 3: that human beings have evolved as protean *superadaptors* who seek, acclimate to, and thrive in wide-ranging physical environments. The Muscle Body (figure 10.5) by the Hyperbody Research Group of the Technical University of Delft (TU Delft) exemplifies the transfigurable environment. Here, a playful, flexible, Lycra envelope reconfigures by McKibben actuators (as does ART's therapy surface) in response to the activities of the "players" inhabiting it.[14] In the LIT ROOM, the transfigurable environment is tuned specifically as a learning environment, where there is ample evidence that changing the environment in which people learn fosters learning by forming new associations in their brains.[15] As Benedict Carey, the author of *How We Learn*, conveys, "The brain wants variation. It wants to move."[16]

The LIT ROOM also acts as a story-extension tool, allowing children to customize environmental effects toward interpreting picture books *for themselves*. In the LIT ROOM, children are invited to become cocreators of their learning environment, which, argues James Paul Gee, professor of literacy studies at Arizona State University, is very motivational to children.[17] In the LIT ROOM, the words of picture books, read by librarians to young readers, literally shape their physical surroundings, giving form to the book. If the atmosphere of a given section of the book, as configured by the environment in lights, sounds, shape, and movements, fails to match the book imagined by the participating children, then these young readers are afforded the opportunity, through the environment's interfaces, to make visible and tangible their own perceptions for inspection by themselves and others. And by the same means, these young readers can alter the course of the book, making visibly and tangibly different outcomes for their engagement and reflection. This active manipulation of the physical environment lends children a sense of ownership and control of their inner thoughts and their external surroundings, each one made to reflect the other. The child becomes invested in the book.

Our belief is that the LIT ROOM will prove effective in cultivating the literacy skills of children in ways that are not only innovative but also well-matched and relevant to the lives of children in a physical and increasingly digital world. These assertions find theoretical validation in the research cited here thus far, as well as in the "cycle of creative imagination" proposed by Vygotsky (the pioneer of developmental psychology), which has also informed HCI work in nonrobotic learning environments.[18] An early exemplar of the transfigurable environment typology of architectural robotics, that "portal to elsewhere," the LIT ROOM project aims to demonstrate that literacy can be advanced within a specially designed public library environment that is at once physical and digital and evocative of the book being read. As will be considered shortly, the LIT ROOM was evaluated by our team with the participation of children in an interactive

10.5
The Muscle Body exemplifies the transfigurable environment, "a portal to elsewhere" (Hyperbody Research Group, 2005; reprinted with permission of the Hyperbody Research Group, TU Delft.)

read-aloud experience of fiction and nonfiction books, led by a literacy expert situated in the Richland Library—the largest public library in South Carolina.

In such a conducive read-aloud environment, *words become worlds*.[19] The LIT ROOM aims to demonstrate that through engagement with a palette of environmental effects (lighting, sound, and movement) at room scale, children's meaning-making can be cultivated within an engaging, mixed-technology environment promoting read-aloud interactions. Key to literacy development, read-aloud interactions are those verbal exchanges occurring between a child and an adult concerning the story being read and its relationship to the child's life.[20] The LIT ROOM serves as both catalyst and capacious home to these exchanges that are so central to literacy development.

THE READ-ALOUD FOR CONSTRUCTING MEANING

If it's not yet evident, the LIT ROOM is a cyber-physical environment dedicated to facilitating what library science calls the *read-aloud*. During an interactive read-aloud, a parent, a teacher, or some other "reading role model" reads to one or more children in order to bond, explain, arouse curiosity, build vocabulary, develop comprehension, and, overall, inspire reading.[21] As opposed to solitary reading, the read-aloud provides a "lived through" experience in which the participating child makes personal, sensory, and emotional connections to the text, which facilitates comprehension and understanding.[22]

Read-alouds in structured learning environments, such as the classroom and public library, are particularly critical for early literacy attainment.[23] As a primary mission of public libraries is the promotion of literacy skills, public libraries commonly offer read-aloud sessions to the children of the communities they serve.[24] For children with limited access to early literacy materials at home, the read-aloud is especially critical, as children reared in poverty tend to be at risk with respect to literacy development.[25]

The read-aloud is an especially rich opportunity for young children to advance language and literacy development when accompanied by interactions with adults and peers.[26] Moreover, studies have shown that the picture book is an especially productive vehicle for read-alouds because of their inherent balance between text and image, providing a rich opportunity for young children to contextualize verbal language in a way that enhances learning.[27] Similarly, there is ample evidence that language growth and school achievement are directly related to early picture-book reading for young children.[28] In sum, the most favorable conditions for a read-aloud experience involves picture books and multiple peers in a physical environment conducive to learning, where the adult reader facilitates interaction.

Recognizing the rapid advancements in ICT, public libraries have increased their efforts to advance literacy skills in innovative ways that satisfy expanding definitions of literacy.[29] Just as picture books promote the contextualization of language by using visual images, advancements in ICT have expanded traditional conceptions of literacy to include the space beyond the printed page of the book. At public libraries, however, the role of ICT in cultivating literacy has primarily focused on providing public access computers connected to the Internet and loaded with screen-based, interactive applications. But interacting with a "keyboard-mouse-screen" interface does not offer the immediacy of interacting with the printed page, and both forms of interaction, the printed page and the computer, are far removed from the physical, tangible, social world that young children thrive in. The divide between cyberspace and the physical world, in particular, leaves all of us, notes Ishii, "torn between these parallel but disjointed spaces."[30] Consequently, in developing the LIT ROOM, we drew inspiration from the *situational* theory of literacy that places equal value on learning tools (books, devices, or objects) and on the physical and social environments within which individuals leverage those tools.[31]

As ICT has expanded the context of learning into a richer variety of domains, a wide range of research has emerged to study how ICT-enhanced learning promotes literacy skills and motivates children.[32] Related more precisely to the LIT ROOM, a segment of this literature is focused on how ICT-enhanced learning is supported through, specifically, interactive reading aloud of picture books. Using different sensing and interaction techniques, researchers have developed *augmented* and *mixed-reality* printed and e-books containing audio and video media, aimed at both expanding and enhancing the reading experience.[33] Other researchers have explored the benefit of combining printed books and e-books with personal computers, tangible interfaces, responsive toys, robots, and augmented-reality viewing devices to enhance the interaction between book and reader.[34] "Story room" environments have utilized technology at the scale of the room to encourage authorship and collaboration between children through storytelling.[35] In merging physical and virtual technologies to create enhanced digital-physical environments, these rooms leverage collaborative learning of narrative construction through a range of shared input devices so children can construct, record, and play back stories. These various state-of-the-art approaches to technology-enhanced, contextual, literacy education have aimed to merge the conceptual space of the book and the real space inhabited by the reader. Compared to the LIT ROOM, however, these prior efforts are limited by their physical size and scale, by the specific technologies they use, and by the range of interactions, especially social ones, they afford.

In augmented and mixed-reality books—books incorporating digital and multimedia elements—the physical scale of engagement is limited to the narrow frame of the reader's immediate reading environment and focus: augmentation (i.e., three-dimensional transformations) occurs only virtually and is dependent on the use of cumbersome goggles or handheld devices. One of these, the Listen Reader, was designed to create an immersive environment for a child reading a picture book through high-quality embedded audio, triggered by sensing technology within the book.[36] Another example, the Wonderbook, a large-format augmented book, combines a physical book with sound and video projections that contextualize the text and provide opportunities for interaction between readers.[37] These two investigations concluded that augmented books can encourage social and collaborative reading experiences, resulting in increased interest in the book being read. But while there was evidence that participants engaged in the Wonderbook experience had recalled critical aspects of the story (particularly those aspects contextualized through sound and video), the study was not yet focused on how readers might be leveraging the multimedia content for meaning-making. In any case, augmented and mixed-reality books, both printed and electronic, may exhibit sensing, interactive, and environmental characteristics, but they are not yet particularly tangible, spatial, or reconfigurable and not especially social.

Other research efforts strive to cultivate literacy in young readers by integrating tangible interfaces into the reading experience. One example, SAGE, a cyber-physical story-listening and authoring tool, combines an interactive plush toy that elicits from young readers unique storytelling strategies that are digitally recorded for playback.[38] When compared to screen-based digital interfaces, the tangible and embodied augmentation of SAGE supported deeper explorations of identity and communication and provided opportunities for children to project fears, feelings, and interests through stories. The LIT ROOM aimed to build upon the promise of cyber-physical literacy tools like SAGE by expanding multisensory feedback and prompting multisensorial effects into the physical environment occupied by adults and children as they engage in shared picture-book reading.

For "story rooms" that merge physical and virtual technologies at room scale, their immersive impact (their interactive imagery) is achieved mostly through audio and cinematographic effects within an otherwise fixed environment. Both the KidsRoom and the Island of Sneetches guide children through a narrative story structure, leveraging physical props and multimedia effects to elicit reactions and responses from the children.[39] Similarly, the interactive StoryMat, upon which children create and record stories using small toys and projected images as prompts, leverages aspects of a room-scaled environment to foster associations within stories and extend their meaning.[40] While studies of children and adults interacting with these systems are mostly focused on usability, some of these studies begin to suggest the promise of room-scaled engagement for providing feedback, fostering collaboration, and eliciting unique responses to the text being read.

Expanding beyond the physical pages of the book, the ICT initiatives considered here—augmented books, tangible interfaces, responsive plush toys, and story rooms—create an engaging, immersive, and interactive reading experience for children. While these state-of-the-art approaches and an ample literature on literacy, learning environments, and the growing influence of technology on children are compelling, an artifact and associated literature has yet to emerge on the important effects of the interactions in these domains. Additionally, there are significant gaps in the literature on effective interventions in children's library environments, particularly where interactive technologies and embodied interaction are the basis for early literacy development.[41] Our research in child-robot interaction sought to fill these gaps by focusing on the design and evaluation of a playful, interactive experience that, in sum, involves:

- A mixed-media environment that is both digital and physical, facilitating a broad range of children's responses to the text.[42]
- The accommodation of multiple participants that encourage dialogical interactions—that verbal exchange across children and adult facilitators that has been shown to advance language and literacy development.[43]
- A palette of lights, sounds, colors, forms, and physical movement to help children leverage mental imagery toward constructing meaning and making sense of the stories being read.[44]
- The ability for children to program this palette for themselves—to construct meaning, analyze, and think analogically and metaphorically through a different sign system.[45]

Transforming a part of the library's larger interior into a robotics-embedded physical environment for children's meaning-making, the LIT ROOM acts as a portal between the everyday environment and the extraordinary environment imagined in books, providing unique opportunities for children and their adult guides within an interactive read-aloud experience, so critical to cultivating literacy.

A SCENARIO: 7- AND 8-YEAR OLD CHILDREN DISCOVER THE LIT ROOM

The following scenario describes how the LIT ROOM transforms an interactive read-aloud into a multimedia, mixed-reality experience in a public library. In this scenario, a small group of 7- and 8-year-olds and a librarian interact with the LIT ROOM, which uses a tablet computer mounted to a small lazy-Susan table to control the atmosphere of the multimedia, architectural robotic artifact. (Referring to the illustrations of the LIT ROOM earlier in this chapter will help contextualize the scenario.) The LIT ROOM functions to provide environmental feedback to read-aloud prompts, contextualizes ideas from the picture book being read, and serves as a tool for a child to make visible her thoughts and ideas. One spring afternoon, a group of six second-grade children, 7 and 8 years old, visit the public library for its weekly read-aloud activity. The group of young children is intrigued by the unexpected: an installation within the children's reading room that looks to be the size of a generous bedroom. At first, this peculiar physical presence appears as a sizable, inhabitable cube, gently glowing, and delivering just-audible ambient sounds. A mostly open, aluminum frame defines the cubic form of the room-scaled environment. The frame is fitted with four curious panels, identical in size, each a bit taller and also far wider than a second-grade student. For three of these panels, each is fitted within one wall frame oriented vertically. The fourth panel is fitted within the ceiling frame. As the children approach the room, they are themselves illuminated by the soft-colored glow emanating from the panels. All four of the panels appear to be floating in mid-air, physically independent of the cubic, metal framework supporting them.

 The floor under and around the artifact is covered by a colorful rectangular mat, far bigger than the footprint of the cube resting on it, and at a small angle to it. The mat seems to identify a larger, physical precinct for the cube, and when the second graders are collectively standing on it, it's as if they have arrived *somewhere else*, in some space outside the library interior. It feels to the children, standing within this precinct defined by the colorful mat, that something special, something rather extraordinary will happen in this strange space.

 Terry, a library staff member with expertise in read-alouds, greets the children as he typically does when they arrive for their weekly read-aloud session, only this time, he directs a group of six of them, initially, to seat themselves within the modest-scaled rectangular volume, around a small, circular table, not more than 16 inches from the floor, located at the center of the installation. Once the children, looking about curiously, settle, Terry takes an empty spot about the circular table for himself. Greeting the children with a gentle, "Hi, welcome," Terry takes the edge of the small table with his hand and gives it a gentle spin, so that all of the children can see that the table is equipped with a small touch-screen display, a typical tablet computer that changes colors and repeats to the children, "Hi, welcome" in Terry's voice as it slowly rotates with the table. The touch display affords direct interaction by the children and by Terry with the books that will soon be read and with the larger, cyber-physical LIT ROOM environment that surrounds them.

Now that the children and Terry are seated attentively about the table, Terry begins to familiarize the children with the LIT ROOM rather than launching prematurely into the read-aloud with printed books. Terry recognizes that if children new to the LIT ROOM have a few moments to explore its capabilities, the novelty of this strange environment is far less a distraction for the young readers. Terry spins the small round table gently again, and it stops so that the display is addressing Maya, a 7-year-old girl from the group. Maya deciphers easily that the tablet displays the four seasons—summer, fall, winter, and spring—labeled by name and by iconic representation (say, a leafless tree for winter). Terry allows Maya a chance to make this recognition and encourages her to speak out-loud about her discovery. Subsequently, Terry asks another child, Adam, to reveal to the group his favorite season. When Adam offers "summer," Terry asks Maya, who has control of the tablet, to select the word *summer* from the tablet options. In the short time since they've entered the library, the children have been transported from its children's reading room to a strange, cubic environment; and now, with Maya's selection from the tablet following Adam's suggestion, to someplace outdoors and warm, where a summer storm has formed. The LIT ROOM's lighting dims and flickers in bright white to evoke thunder, rumbling sounds are heard, and the LIT ROOM's panels move joltingly to suggest flags, banners, or sails in the wind. The unease of this summer storm evoked by the LIT ROOM delights the children. They hardly recall that, outside the library, it's a remarkably calm spring day, and the flowers and trees are only budding in its sunshine.

Terry poses to the children a key question: Is this the summer storm they had imagined or do they imagine a summer storm to be different than the LIT ROOM has presented it? Terry looks at Daniela specifically, asking her what she thinks. She says that she can recall a summer storm in her native Mexico that has more red in the sky and more breeze than wind. Terry asks Maya to spin the small table so that Daniela can access the tablet. Daniela recognizes that the tablet's display is now organized differently. In columns, icons are dedicated to different sounds, colors, forms, and movements that, presumably, the LIT ROOM can assume. Daniela, who has control of the tablet, is encouraged to create the red color and relative stillness of a summer storm in her mind, and the LIT ROOM follows her commands, assuming her selections, becoming *her* summer storm. Terry tells Daniela and the rest of the children that they can "save" their selected preferences for later recall, and that the LIT ROOM can also take a photograph of the environment they have created and print this to take home. With Terry's guidance, the children try a few more first engagements with the LIT ROOM, exploring how the system evokes an emotion, a temperature, and, more abstractly, two qualities of surfaces—jagged and smooth.

Now that the children seem comfortable and reasonably acquainted with the LIT ROOM, Terry taps the touch screen and opens the standard, unmodified, print version of the children's picture book, *Underground*.[46] A book of historical fiction, *Underground* uses evocative images and language to chronicle the journey of a group of North American slaves from captivity to freedom on the Underground Railroad. Terry introduces the children to the book and to how the strange LIT ROOM that surrounds them might bring to life its pages:

> This book, the *Underground*, is based on a true story of how people who were slaves escaped captivity and, with the help of other people, gained their freedom. As we read the book, the LIT ROOM will transform itself to represent the emotional quality of the three main sections of the book—*escape*, *help*, and *freedom*—using motion, light, and sound. If you look, now, at the computer tablet, you will find displayed there a picture of *Underground*'s book cover (see figure 10.6, far left, as displayed on our earlier, tangible cube interface rather than on the current table interface shown in figure 10.4). Maya, please touch the screen to reveal images from the main sections of the book: one expressing *escape*, one depicting *help*, and the third illustrating *freedom*. These images on the screen will guide our interaction with the LIT ROOM during our reading experience today.

As Terry has finished explaining the overall concept of the book and how the LIT ROOM will support its reading, he begins reading *Underground* aloud to the group of children, again, using only his printed copy.

10.6

Matching words and images using a tangible cube interface. This tangible cube interface was subsequently replaced with the current touch-screen tablet shown in figure 10.4 for both the LIT ROOM and the LIT KIT, its compact outreach component.

When Terry arrives at the part of the book where the characters begin their escape, he prompts Sam, an 8-year-old girl, to select from the various tablet options the image that represents the concept of *escape*. The correct image directly correlates with the illustration in the book. After Sam selects the correct image, the tablet changes to depict an image of running feet from the book and two words that express emotions. After reading a bit more from *Underground*, Terry asks another child, Henry, to consider what it might feel like to be running away from something he was trying to escape. Terry prompts Henry, who is now in command of the tablet, to pair the image of running feet with the word that best expresses that feeling. At Terry's prompting, Henry correctly pairs the image of running feet with the word *afraid* (figure 10.6, far right, as displayed on our earlier, tangible cube interface again). With Henry's correct pairing of icons, the LIT ROOM provides environmental feedback: its lights alternate bright-blue and red, and its robotic panels undulate to create a cluster of fast, chaotic movements representing a group of people moving together. Upon hearing the sound of people running through a forest, another child, Jenna, exclaims, "the shadows look like images of people running—the sounds make me think of times I've felt scared, too." Terry continues reading, repeating the same process for the concepts of *help* and *freedom*.

After reading through the book once, Terry and the children discuss its major themes and decide to customize the environmental effects for each section by selecting the "Set-Up" menu from the tablet. As Daniela selects colors, sounds, and movements for the first section of the book, Terry asks her to explain her choices, providing an opportunity for her to leverage the LIT ROOM environmental effects to create deeper meanings and connections to the story. Accordingly, Adam and Jenna fine-tune the environment to evoke what they imagine to be the other two sections of the book, and Terry follows again with more discussion.

In their thirty minutes together, Terry and the children discovered that the LIT ROOM's programmed configurations sometimes matched a child's imagined space and other times provided the opportunity for the children to fine-tune the room through word-image couplets. In their final minutes together, Terry prompted the children to change the course of the book electively by reprogramming the environment: to explore different possibilities and outcomes (as in *What-if?*) to make visible their thoughts and desires.

THINKING BIG BY (FIRST) WORKING SMALL

Given the larger physical scale and complexity of the LIT ROOM, we decided to start much more modestly, physically and technically. But we didn't want the scaling down of this artifact to scale down our ambition. In our early research efforts, we sought a happy medium (you might say) that retained the capacity to explore child-machine interactions that test our hypothesis: that literacy skills can be advanced in an environment that is physical, digital, and technologically rich. We imagined that this more modest artifact could nevertheless be architectural robotic, a reconfigurable room pulled out of a box that could easily be transported by a single person. We also wanted the artifact to be built (by us, at least) at a total cost of less than $300 excluding labor. We figured that if we could familiarize ourselves with how children interact with this modest prototype, we would understand the larger problem and ambition far better, which would inform continued development of the room-scaled LIT ROOM. Additionally, we envisioned this compact LIT KIT to serve, ultimately, as an outreach component to the larger, room-scaled LIT ROOM environment occupying the library. In this way, the LIT KIT extends the occasional library experience of the LIT ROOM to a prolonged engagement with this technologically rich environment in, also, homes and classrooms. *We built small to think big.*

    Like the LIT ROOM, the LIT KIT (figures 10.7, 10.8, and 10.9) uses color, sound, and movement to create an environment that is evocative of the picture book being read. Easily transported, the LIT KIT (16 inches wide by 16 inches deep by 30 inches high) can be carried by its comfortable handle by one person using one arm. The box has a cover that is unlatched and pulled off when in use. On each of the box's lower exterior walls is mounted, flush with the surface, lighting that can project a wide range of colors on neighboring walls and other surfaces. Also flush-mounted on the box are audio speakers. Mounted on the top surface of the open box are four panels, each one actuated by a servomotor. Attached to each panel is an 8-foot length of cloth ribbon terminated by a suction cup and a hook for mounting on the wall, ceiling, furnishings, or otherwise. Combined with the colored lights and the sounds emanating from the speaker, these extended ribbons, actuated by the servo-driven panels, create the room-scaled, environmental effects of the LIT KIT. The upper-half of the LIT KIT's cubic volume is fitted with a drawer to store several picture books and the instructions for the LIT KIT's use (a design problem, itself). Prior to using the larger tablet computer interface as described in the scenario for the LIT ROOM, we designed a Sifteo cube interface for children to access the LIT KIT's room-scaled environmental effects (see figure 10.6).[47] At 1.5 inches on a square side, and a third that in depth, the Sifteo cubes are easily stored in the LIT KIT drawer along with some picture books and the instructions. A standard Arduino board, housed within in the lower portion of the LIT KIT with the lighting and speakers, serves as the computer for the system.[48]

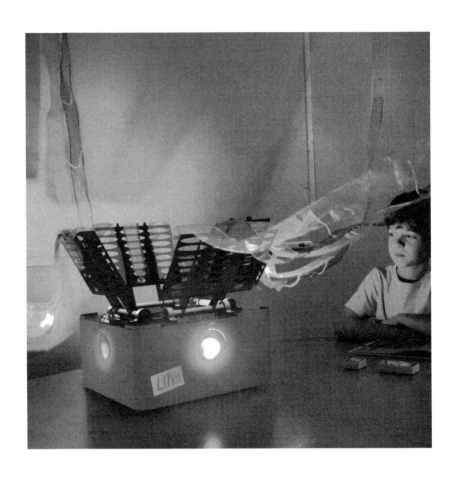

10.7
The LIT KIT final prototype (photograph by the author).

10.8

A student, a reader, a printed *Underground* book, and the LIT KIT—a portable LIT ROOM for home and classroom (photograph by the author).

10.9

The LIT KIT in a classroom filled with children (photograph by the author).

A LIT ROOM for the Persistence of Print

PUTTING THE LIT KIT TO THE (CLASSROOM) TEST

In use, the LIT KIT captures the essence of the LIT ROOM experience. First, children use a graphical user interface (GUI) for capturing their input as the story unfolds. Second, the room transforms in a palette of colored lights, sounds, forms, and movements to evoke the book being read. And third, children have the opportunity to program for themselves an evocative environment that reflects *their* ideas and concepts inspired by the picture book and its consideration by the larger group. Accordingly, use of the LIT KIT is described well by the scenario for the LIT ROOM considered earlier, the key difference being the scale, complexity, and richness of the interactive experience.

Given that the LIT KIT provides a good facsimile of the LIT ROOM experience, we learned a lot, early on, about our larger research ambition by conducting an empirical study using the first fully functioning LIT KIT prototype. The research questions we addressed are ones common to interacting with the artifact at either the KIT or ROOM scales:

- Can children and adult readers comfortably use the technology during interactive read-alouds?
- How do the environmental effects aid users in the creation of meaning?
- Can the LIT KIT's effects contextualize language, concepts, and ideas in different types of picture books?

The evaluation of the early LIT KIT prototype (see figures 10.7–10.9) was carried out at a public elementary school over two days. Participants in the study included teachers and students from two second-grade classrooms (participating children were ages 7 to 8, except for one child in classroom "B" who was 11 years old). The LIT KIT was evaluated during whole-class, read-aloud sessions as well as in one-to-one (one child, one adult reader) read-aloud sessions. For the whole-class sessions, we had twenty-three students participating in classroom "A" and twenty-four students in classroom "B." For the one-to-one read-aloud sessions, we had seven boys and seven girls across the two classrooms participate in our investigation. Two picture books with different characteristics were selected for the sessions. Relative to the research questions posed, the evaluative sessions were designed to elicit data on the system's *usability* (following from the first question offered in the bulleted list above), effectiveness as a tool for *meaning-making* (following from the second question), and its *applicability* across different types of picture books (following from the third question).

To gather data on whether the participating teachers and children could comfortably use the LIT KIT during and after an interactive read-aloud, the research team was guided by established protocols for evaluating interactive technology with children, as offered by Read and Markopoulos.[49] Initially, users interacted with the LIT KIT during a whole-class session, with the classroom teacher facilitating the read-aloud and administering the prompts for one of the two picture books. The research team observed the read-aloud, making notes on how the users engaged with the Sifteo cube interface and how users reacted to the overall system. Immediately after the whole-class, read-aloud session, the teacher and the research team asked the children for feedback about their experiences reading the book with the LIT KIT. The session was audiotaped and videotaped, and the discourse was subsequently transcribed. Additionally, the teacher of the class completed a questionnaire that evaluated the LIT KIT on measures of usability and appropriateness for use in a read-aloud activity.

After the whole-class sessions, seven children, chosen by their teacher to represent a diversity of learner characteristics, participated in the one-to-one sessions in a private reading room. For these sessions, the second picture book was read aloud by a member of our research team with expertise in interactive read-alouds (see figure 10.8).

Participants were prompted to interact with the LIT KIT during the read-aloud, followed by a customization exercise where each child programmed the environmental effects to represent her or his own mental image of one specific concept from the book. The research team notated its observations, and the sessions were audiotaped, videotaped, and observed for subsequent transcription of the discourse and verification of our notes.

Participants then completed a facilitated questionnaire that rated the prototype on measures of usability, environmental impact, aesthetic design, reading motivation, and engagement. For these one-to-one sessions, the research team used an evaluative questionnaire designed specifically for children ages 6 to 8. The specific instrument used in the questionnaire was the Smileyometer, a graphic representation of the traditional Likert scale, where a numbered scale of, say, 1 to 5 (registering a response of less to more, harder to easier, and so on) is replaced by faces (frowning-smiling) that illustrate varying levels of approval (figure 10.10).[50]

Questions in the questionnaire included the following:

- How easy was the LIT KIT to operate?
- How easy would it be to explain how to use the LIT KIT to one of your friends?
- How much did the LIT KIT change the room's environment?
- How much did you like the way the LIT KIT changed the room's environment?
- How much did the LIT KIT feel like a game or a toy?

- What did you think about the LIT KIT's packaging?
- What did you think about how the LIT KIT system looked when it was assembled?
- How much did the LIT KIT help you to understand the book?
- How much did the LIT KIT make reading the book fun?
- How much would you want to use the LIT KIT to read another book?

The questionnaire also used open questions for which participants provided verbal explanations. Open questions in the questionnaire included the following:

- Which of the LIT KIT's environmental effects did you like the most, and why?
- What other types of things would you like to see the LIT KIT do?
- Do you have anything else you'd like to share with us about your LIT KIT experience today?

To offset any potential limitations of the Smileyometer as a research instrument, such as the inclusion of leading or confusing questions, results of the questionnaire were confirmed against documented observations of usability errors exhibited by the children. In any case, it is important to recognize that in performing usability and efficacy studies with child participants, researchers must be cognizant of the developmental capabilities of the age group being studied and recognize that the results of such studies are relatively more suggestive than when working with adult participants.[51] For one, more than adults, younger children completing a Likert scale (here, the Smileyometer) may occasionally circle two indicators (two of the faces on the scale) rather than one face or may not correctly correlate a given face to the intended response.

10.10
The Smileyometer from our questionnaire
(diagram by the author).

The outcomes of such qualitative studies with children must therefore be taken as more suggestive than the outcomes from like studies with adults that, anyhow, are not definitive but useful as a guide to forward the iterative design process.[52]

To explore how the LIT KIT contextualizes language, concepts, and ideas in different types of picture books, two very different books were selected for the study by an assembled group of education experts. One of these was the narrative, nonfiction picture book *Four Seasons Make A Year*,[53] which teaches children about the changing seasons; the second book was the historical-fiction picture book *Underground*, introduced earlier in this chapter within the LIT ROOM scenario. Children's responses were analyzed to suggest how the LIT KIT's environmental effects (colored lights, movements, sounds) might evoke not only the physical phenomena (trees, wind, rain, etc.) presented in *Four Seasons Make A Year* but also the emotions (fear, sadness, happiness, etc.) evoked by *Underground*. The books were counterbalanced for the two classrooms. Classroom "A" read *Underground* together in the whole-class setting, followed by *Four Seasons Make A Year* in the one-to-one sessions, whereas classroom "B" read the books in the opposite order. Additionally, we asked the teacher in each class, by questionnaire (Likert scale and open questions, again), to rate the LIT KIT on its applicability to a variety of picture books.

HOW THE LIT KIT TESTED IN THE CLASSROOM

Feedback from the participants in both the whole-class and one-to-one read-aloud sessions suggested that the system was easy to use, provided effective feedback to vocabulary and comprehension-based prompts, made reading the books fun and engaging, and created an environment that helped the children visualize concepts and think about the book. Additionally, we found that children and teachers gave an equally positive evaluation of the LIT KIT, irrespective of which of the two books was being read.

On Usability

The two participating teachers reported that the LIT KIT was easily integrated into their read-aloud sessions with minimal training on the system. Based on a post read-aloud survey, both teachers indicated that the LIT KIT was easy to use; an appropriate and useful tool for elementary literacy education; and made the interactive read-aloud fun and engaging for students. Finally, the teachers offered that they would use the LIT KIT again in their classrooms.

During the whole-class read-aloud sessions, we observed problems encountered by some children in their interactions with the Sifteo cubes, whereby a few participants continued pressing the cubes multiple times or, alternatively, pressed the cubes too gently to register a response or, again, failed to place the cubes side by side when this was the expected response. The tablet computer interface used for the LIT ROOM (see figure 10.4) is, however, more familiar to children than Sifteo cubes, which overcomes this obstacle of the current LIT KIT prototype. In future implementations of the LIT KIT as an outreach mechanism, the research team will use a small table in place of the Sifteo cubes.

For the one-to-one session, the questionnaire data suggest a positive rating from children for the LIT KIT on measures of usability, aesthetic design, environmental impact, and reading motivation. High ratings on reading motivation suggest that the children surveyed would use the LIT KIT again for a read-aloud, and that the system may have helped them understand the book better through an engaging and fun experience. Further reporting on and consideration of usability outcomes is found in a technical paper published by our team.[54]

On Meaning-Making

As picture-book read-alouds are discursive events where learners construct meaning together, one important method for understanding children's meaning-making is to document and analyze children's response initiations (i.e., their verbal talk).[55] The whole-class and the one-to-one sessions were transcribed, noting instances where children seemed to be leveraging the LIT KIT to construct meaning related to the picture books being read. Furthermore—and guided by Baer's research into children's response to texts using visuo-spatial activities—the research team conducted an initial coding of the verbal responses of participants engaged in the LIT KIT studies in order to suggest specific meaning-making strategies.[56] Additionally, the questionnaire completed by the teachers after the whole-class read-aloud session asked them to rate the LIT KIT on its effectiveness as a tool to scaffold meaning-making.

The teachers reported that the system was a useful tool for scaffolding meaning-making and vocabulary acquisition in the context of an interactive read-aloud. During the whole-class sessions, children spontaneously responded to the LIT KIT's environmental effects and in many cases related them directly to the book being read. A child in classroom "A," experiencing the effects that contextualized the season of fall, said: "Cool! That looks like leaves on a fall tree!" Children in classroom "B," during the reading of *Underground*, shared multiple interpretations of the KIT's "escape" effects. Looking at the lighting and shadows cast on walls and ceiling, one child commented: "They were walking hand in hand," while another remarked, "Those are footprints!" These kinds of responses indicate that children are able to consider both concrete objects (trees, seasons, and people) and emotions (such as fear) while immersed in the multisensory environment created by the LIT KIT. Additionally, the research team observed children spontaneously assisting and collaborating with each other to interpret and relate the environmental effects to the picture book being read. In one instance, a child looked puzzled and voiced "I don't get it," which prompted a neighboring child to help the puzzled child by pointing to the shadows and suggesting, "It was the people running. I can see that!"

Many children in the whole-class setting indicated that the LIT KIT was an effective tool for helping them visualize the books being read. After the read-aloud, the research team asked the open-ended question: Do you have any comments about the KIT that you would like to share? One child stated: "I think [the LIT KIT] was really cool and fun, and I actually think it would help with the pictures from the book, because sometimes when a teacher's reading to me, I want to be able to see the pictures in my head and [the LIT KIT] really helps." Many comments supported the capacity of the sound effects, in particular, to help the children understand what was happening in the books. Other children commented positively on the feedback component of the system, describing the Sifteo cube prompts as both "fun" and "challenging."

An initial analysis of the children's verbal response initiations, both in whole-class and individual picture-book read-aloud sessions, revealed many instances where the LIT KIT supported meaning-making processes. Children were able to leverage the palette of environmental effects to make visible their imaginings and ideas. The lighting, movement, and sound, rather than distracting children from key constructs in the texts, formed an effective palette for elaborating meaning. Moreover, children were able to articulate verbally the reasoning behind their choices, suggesting that the multimedia, mixed-reality elements helped them explain *how* they constructed meaning.

Overall, the LIT KIT elicited a wide range of response initiations from the participants in the study. Some children constructed meaning by using the multimedia effects to help them make sense of the picture book being read (table 10.1, A), while others used them to make connections between the text and their own life experiences (table 10.1, B). Students also created meaning by articulating the feelings or physical sensations depicted in the picture book (table 10.1, C), while others used the environmental effects to elaborate on the story (table 10.1, D).

On Applicability across Picture-Book Types

The teachers indicated that the LIT KIT could be adapted to a variety of picture books. We observed the young participants using the multimedia LIT KIT effects to represent the physical phenomena (trees, wind, rain, etc.) and actions (swimming, eating, flying, etc.) evoked by the *Four Seasons Make A Year* picture book. For *Underground*, children successfully transformed the atmosphere of the schoolroom with the LIT KIT to evoke the feelings of the characters (fear, sadness, happiness, etc.) and abstract concepts (such as slavery and freedom).

TABLE 10.1

LIT KIT "RESPONSE INITIATIONS," WITH EXAMPLES FROM CHILD PARTICIPANTS

A. CHILDREN MAKE SENSE OF THE TEXT.
Comparing how the LIT KIT changed from "spring" to "fall" for *Four Seasons Make a Year*, one participant said: "The light and motion are changing because the leaves used to be green, and now they changed to orange. And the breeze is calmer in summer than in fall."
B. CHILDREN MAKE CONNECTIONS BETWEEN THE TEXT AND THEIR OWN LIFE EXPERIENCES.
In response to the LIT KIT spatializing the "fall" season in *Four Seasons Make a Year*, one participant said: "The light and shadows make me think of when it is fall: me and my brother, we'd be making a big pile of leaves and be jumping in it. And then the leaves go everywhere."
C. CHILDREN SHOW EMPATHY BY UNDERSTANDING FEELINGS OR PHYSICAL SENSATIONS DEPICTED IN THE TEXT.
In response to the LIT KIT evoking the concept of "freedom" in *Underground*, one participant said: "It makes me feel good and happy because I play basketball and it looks and sounds like people cheering, and like, congratulating people for being free."
D. CHILDREN ELABORATE ON THE TEXT.
While explaining her rationale for selecting the *twist* motion-setting to represent the concept of "escape" in *Underground*, one participant said: "When wind twists, it makes a tornado, which is scary like this."

After children explored the custom settings of the LIT KIT, the adult reader had a dialogue with the children to discern their rationale (see table 10.1). To contextualize the concept of "spring" from the book *Four Seasons Make a Year*, children chose a variety of lighting effects and discussed their reasoning (*green* to represent leaves; *blue* to represent rain, *red* to represent apples and flowers), sounds (*wind and rain; eating an apple* from an apple tree; *running* because "when it rains, people run to avoid getting wet"), and movements (*rotate* and *sway* to represent the motion of wind in the trees; *undulate* to represent the motion of "someone swimming"; *soar* to represent the motion of a "bird's wings in the air"). In many of these cases, children were constructing ideas about "spring" that went beyond what was represented in the picture books, making connections to other texts and their own knowledge base and experience.

The LIT KIT was equally effective at contextualizing an abstract concept such as "escape" or the emotional qualities evoked by the *Underground* (see table 10.1). Children chose a variety of lighting effects (*red* to represent anger, fear, and nervousness; *blue* to represent the color of night and the emotion of sadness), sounds (*wind*, *running*, and *nighttime* sounds, each directly related to the context in the book), and movements (*shake* to represent the motion of the slaves "shaking with fear"; *sway* to represent the ominous sounds of night that evoke fear; *twist* for the "slaves stumbling over each other as they ran"). Many children used the multimedia effects to extend the meaning in the book, thinking both analogically and metaphorically, which suggests high-order comprehension.

IN THE WORDS OF KIDS

The children's open responses to the LIT KIT also validated our intent in more visceral terms. For instance, a child in classroom "A," experiencing the LIT KIT contextualizing autumn, said: "Cool! That looks like leaves on a fall tree!" Meanwhile, during the reading of *Underground,* children in classroom "B" shared multiple interpretations of the LIT KIT's evocation of "escape," offering such remarks as, "They were walking hand in hand" and "Those are footprints!" These kinds of responses suggest that children are able to consider both concrete objects (trees, seasons, and people) and emotions (such as fear) while immersed in the multisensory environment of the LIT KIT. Additionally, the research team observed children spontaneously assisting and collaborating with each other to interpret the LIT KIT's environmental effects, and then relating these effects to the picture book being read. This was evident in the case considered earlier in which a child, puzzled, was assisted by a neighboring child. The remark of another participant, cited earlier, captures the spirit of our ambition for the LIT KIT: "When a teacher's reading to me, I want to be able to see the pictures in my head, and the LIT KIT really helps."

WHAT WE LEARNED IN SCHOOL

We learned a few things by bringing the LIT KIT to school and the LIT ROOM to the Children's Reading Room at Richland Library. We learned that the LIT ROOM and LIT KIT cultivate reading interactions across participating children and the adult reader-facilitator in a read-aloud setting. We learned that the two architectural robotic read-aloud environments leverage multimedia effects to provide feedback, prompt children's responses to the text, and create unique environments for them to think about different types of picture books. We learned that both reader-facilitators and children find the LIT ROOM and LIT KIT easy to use, fun and engaging, as well as helpful to their understanding of the books read. We also learned from adult reader-facilitators that the LIT ROOM and LIT KIT create unique opportunities during and after the interactive read-aloud for them to fulfill their role as guides, facilitators, questioners, and responders—a critical factor for children's meaning-making.[57]

We also learned from the children who interacted with the LIT ROOM and LIT KIT something more central about the transfigurable environment typology of architectural robotics—that *portal to elsewhere*. Of those children who engaged the LIT ROOM and LIT KIT, a good number of them, in their verbal responses to our open-ended questions, focused their commentary on the evocative atmospheres created by these transfigurable environments: on how the sounds, colored lights, forms, and movements connected them to the book, supported their understanding of it, and helped them contextualize or generalize this understanding to broader themes in their own lives. Validating our aim, the sentiments of the children are in turn supported by Gambrell and Bales' theory of children's meaning-making. This theory suggests that when connections are made to a picture book through visuo-spatial activities, children use mental imagery to mediate between an idea about the text and an understanding of the world at large.[58] A child first makes an abstract mental image of some concept about the picture book, translates it into a concrete, spatial representation (in this case, through environmental effects), and then, through discussion, makes connections between the idea, its representation, and the world.

What has been described here is a complicated journey: one not well understood, and yet, one central to our understanding of ourselves and the world around us. From this broader perspective, much of what we learned from the LIT ROOM and LIT KIT research is supported by a recent study by Kathryn Hirsh-Pasek, professor of psychology at Temple University, which finds that language skills are less dependent on the quantity of words a child hears and far more dependent on what children and adults do with these words.[59] Full of evocative sounds, colored lights, forms, and movement, the LIT ROOM and LIT KIT provide a cyber-physical environment meant to encourage children to dwell longer and deeper inside the picture book being read. "Inhabiting the book" in this way prompts the exchange of words across children and the adult reader on matters stemming from the book—issues, fears, concepts, ideas, and aspirations—that is arguably the most convincing demonstration of meaning-making and literacy.

For young readers, the LIT ROOM and LIT KIT provide a conduit between the everyday environment and the extraordinary environment imagined in picture books. Prompted by our research and the theoretical lens of Gambrell and Bales that partly guides it, we arrive at a fundamental question: What is the relationship between the mind and the physical world? This is a question this book returns to often, one that is particularly poignant as we consider the *transfigurable*.

And it is precisely the question Michael Graziano, professor of psychology and neuroscience at Princeton University, poses and then addresses in his recent book, *Consciousness and the Social Brain*.[60] As Graziano argues, we

10.11

White is not white. Claude Monet, *Study of a Figure Outdoors (Facing Left)*, 1886. (Paris, Musée d'Orsay. Michel Monet donated this painting to the Louvre in 1927. Reprinted with permission of Musée d'Orsay, Paris. © F.L.C. / ADAGP, Paris / Artists Rights Society [ARS], New York, 2014.)

know something about the brain and the body, but not much about the internal life inside us, consciousness: that subjective being that makes sense of experience. Drawing from social neuroscience, attention research, control theory, and other fields, Graziano argues that our experience of the world is not as direct as we imagine but instead mediated: that "our brain computes models that are caricatures of real things."[61] To demonstrate this argument, Graziano offers a familiar fact: that we experience white light as "white," even though white light is not white but a mixture of colors of the visible spectrum. The Impressionist painter Claude Monet demonstrated this fact in his *Essai de figure en plein air (vers la gauche)* ("Study of a Figure Outdoors [Facing Left]"): what is clearly a white dress, irradiated by a summer Sun, is painted in every color but white (figure 10.11). All of our experiences, Graziano argues, are like this.

In this murky space between the world and our experience of it, the LIT ROOM and LIT KIT, exemplars of the transfigurable environment typology of architectural robotics, provide a playful, evocative place full of triggers, prompts, and suggestions for young readers to locate themselves *in a place that's theirs*. As a portal to this someplace else, the LIT ROOM and LIT KIT promise, borrowing Antle's words, "to advance a child's grasp of our universe through active, creative exploration."[62] It is not a direct portal to truth, however, but a strange and yet familiar vessel for us to find our way between the world around us and our understanding of it. In a transfigurable environment, we are all Sendak's Max on a far away adventure, right at home.

THE PRIMACY OF PRINT

What should be apparent from the LIT ROOM and the LIT KIT, and from AWE and ART, is that architectural robotics is very much about maintaining the primacy of the physical world in an increasingly digital world through means that are, … well, digital. How strange this is, to develop and embed advanced digital technologies in the built environment in order to return us to the physical world, to maintain the physical world's primacy and realness; this, because the physical world *matters* to us, because it has been our home for 200,000 years. With the LIT ROOM and the LIT KIT, my research team and I have dedicated ourselves to designing, prototyping, and evaluating two cyber-physical artifacts aimed at bringing kids to read books *in print*. Maybe, too, the same two cyber-physical artifacts will inspire some of our readers to one day pursue STEM careers. In any case, the reading experience is our fundamental preoccupation with this research, and the reading experience is one that clearly favors print, a conclusion supported by a wealth of data, summarized by Ferris Jabr in *Scientific American*, as follows:

> Modern screens and e-readers fail to adequately recreate certain tactile experiences of reading on paper that many people miss and, more importantly, prevent people from navigating long texts in an intuitive and satisfying way. In turn, such navigational difficulties may subtly inhibit reading comprehension. Compared with paper, screens may also drain more of our mental resources while we are reading and make it a little harder to remember what we read when we are done.[63]

For Jabr, this means that, "Whether they realize it or not, many people approach computers and tablets with a state of mind less conducive to learning than the one they bring to paper."[64]

This brings us to the question: What do we know, specifically, about children and screen-based reading? The answer is not so much, as there aren't yet longitudinal studies on the effects of tablets and e-books, given the very recent, widespread adoption of these devices.[65] Recognizing this gap in the literature, Julia Parish-Morris and her colleagues conducted what is perhaps the most rigorous studies comparing e-book and printed-book reading on measures of comprehension and also *dialogic interactions* (that verbal exchange between child and adult that is central to literacy development, referred to earlier).[66] With participation of 165 parent-child pairs, the studies by Parish-Morris and colleagues found that children reading traditional books excelled in *higher-level comprehension* (comprehension of the story content and its sequence of events) compared to those reading electronic books and, likewise, engaged in more adult-child, dialogic interaction.

This phenomenon can be explained partly by the observation of Parish-Morris and her colleagues that, with an e-book, parents are busy directing the child's use of the device, rather than posing questions and catalyzing discussion about the story being read. On the contrary, the read-aloud experience with the LIT ROOM and LIT KIT places the printed book *at the center* of the reading activity; the two cyber-physical artifacts are meant not as a book substitutes but rather as evocative settings that prompt and cultivate discussion of printed-book reading.

But there is another, more fundamental explanation for the many benefits provided by the printed book over the electronic book experience: what Jabr refers to as the "reading brain," which, strangely enough, returns us again to the physical environment. Remarkably, the "reading brain," writes Jabr, recognizes text as "a tangible part of the physical world we inhabit."[67] As writing has arrived very recently in the evolutionary history of humans, so the brain "makes do" during a reading experience by working with the neural tissue devoted to related abilities, "such as spoken language, motor coordination and vision."[68] *Text becomes a physical thing.*

And while individual letters are physical things, blocks of text are physical landscapes. "The exact nature of such representations remains unclear," writes Jabr, "but they are likely similar to the mental maps we create of terrain—such as mountains and trails—and of man-made physical spaces, such as apartments and offices."[69] Ah, back to physical spaces, back to architecture, the earliest form of ICT. Printed books are rooms filled with things: we find our way through their letters, sentences, and paragraphs that sit on left and right pages, each page having four corners; and we find our way through printed pages, each book having so many, the heft of them providing us tactile cues as to *where we are*. Accordingly, if someone near to you wants to know *how far along* you are in reading this book, they might ask you *"Where are you* with that book?" recognizing that you are, indeed, *in* the book. *Words become worlds.*

All this paper and print makes navigating our way through printed books easier than navigating screens, e-readers, smartphones, and tablets that deliver what is essentially a weightless media that fades away when you *shut down* or when the device's battery is drained. The physicality and structure of printed media help us form a coherent mental map of our reading experience. Undoubtedly, the very old technology of printed media has many advantages over screen-based reading, and rather than my research team and I joining the wave of ICT efforts to make digital reading more like reading print, or digitally augmenting reading, or making it interactive through digital means, we have dedicated our efforts to focusing on and evoking the words read from printed books through a cyber-physical environment. With the LIT ROOM and LIT KIT, we aim to draw children, using the means of ICT, to reading printed books, to being read to from printed books, and to loving them (figure 10.12).

Meanwhile, our schools and libraries are challenged to meet the call for digital-age skills—including collaboration, creativity, critical thinking, and design—that prepare library users for global citizenry, as mandated by the National Education Technology Plan, the 21st Century Skills policy guidelines, and the standards of the International Society for Technology in Education.[70] Yet, if we look to recent research findings in the cognitive sciences, the "main foundations" for nonverbal and creative reasoning are not the visual stimulation of digital media but "body movements, the ability to touch, feel, manipulate, and build sensory awareness of relationships in the physical world."[71]

My research team and I have no illusions of or nostalgia for returning to some past time and place: that *old* living room with its petrified mantle, fire-a-glow and fireside, a leather upholstered chair and side table, upon which rests the printed tome, the great work, bound with its golden tassel visibly marking where the reader left it (*Fahrenheit 451* redux?). But unmistakably, in reading, there is something important and central to us beyond the two-dimensional surface of a tablet computer. In the LIT ROOM and LIT KIT, we have it both ways and also (and more powerfully) *combined* (digital and physical and cyber-physical) at room scale. These two exemplars of the transfigurable environment typology of architectural robotics, *portals to elsewhere*, are dedicated to empowering children by way of playful interactions between them and adults, the mind and the body, the printed book and a cyber-physical environment that, taken together, help young readers discover the world of reading. As we come to live, play, and work with architectural robotics, my research team and I fundamentally strive to cultivate *in children of all ages* the capacity to engage and shape a world that is increasingly digital, yet innately social and tactile.

10.12
In the LIT ROOM, the environment and its young inhabitants are mutually transformed by an engagement with printed books (photograph by the author).

11 THE TRANSFIGURABLE ENVIRONMENT

The transfigurable environment is that *other place*. And that *other place* may be the dunes on a beach with strong sea breezes, a bit of cool mist, and then, some Sun breaks through fast-moving clouds (figure 11.1). In this way, the transfigurable environment is very much what Negroponte defined as a "simulated environment" where "one can," for example, "imagine a living room that can simulate beaches and mountains."[1]

The evocative forms of the transfigurable environment emerge from the walls, floors, and/or ceilings, as do the functional forms, the furniture-like elements of the reconfigurable environment. However, as the transfigurable environment is not providing useful furnishings but rather evocations of someplace far outside the familiar, the conventional rectangular room it may occupy, along with any semblance of furnishings it may contain, dissolve away. The room is more womb or bladder than bedroom.

An exemplar of the transfigurable environment, the LIT ROOM, embedded in the everyday, physical space of the library, is radically transformed by the words of readers, reading from printed children's books, so that the everyday space of the library *merges* with the imaginary space of the book. *The book is a room; the room is a book.*

In the most optimistic and awe-inspiring way, the transfigurable environment is meant to follow from the well-known Latin saying, *Vulgus vult decipi, ergo decipiatur* ("People want to be deceived, so deceive them"); it invites its inhabitants to be transported from the familiar to the unfamiliar, from reality to illusion. But the distinction between *unfamiliar* and *familiar*, *reality* and *illusion*, is not so clear here. As Albert Einstein understood, "Reality is merely an illusion, albeit a very persistent one."

Unmistakably, the transfigurable environment isn't charged with providing us the comforts of home. Quite the contrary: the transfigurable environment propels us from the comfort zone to uncharted territories, all in one space; because finding ourselves elsewhere is another (strange) form of coming home, as Max does within the confines of his bedroom, inside Sendak's book, after an adventure on the high seas. *The remarkable feature of the transfigurable environment typology is that a single physical space becomes "a portal to elsewhere."*

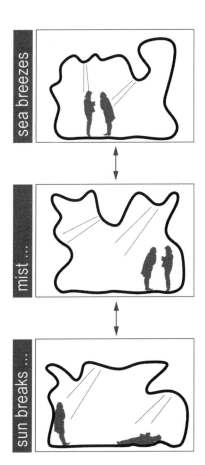

11.1
Diagram of the transfigurable environment typology (diagram by the author).

EPILOGUE

12 ECOSYSTEMS OF BITS, BYTES, AND BIOLOGY

One way to define architectural robotics is to define what's not architectural robotics.[1] It's not *human-robot interaction* with its fixation with humanoids, mostly proprietary ones, programmed to serve people. It's not *intelligent buildings*, focused on temperature control and the opening and closing of windows and shading devices. It's not *intelligent environments*, outfitted with arrays of sensors to capture the everyday activities and mishaps of their inhabitants. It's not *digital fabrication with robots*—industrial robots programmed to manufacture buildings and their components. It's not *buildings with mechanically moving parts*, where walls, floors, roofs, and maybe entire rooms are repositioned. And it's not *human-computer interaction* in its most persistent form: still caught up in the screen.

Closer to architectural robotics are *tangible computing* and *physical computing*, which capture a mix of the digital and the physical (at the sacrifice of the human, at least by name). Closer yet to architectural robotics, perhaps, is *cyber-physical systems*, where we retain the digital-physical mix and gain the systematic relationships forged between constituent parts (and, possibly, the relationship of one of these systems to other systems).[2] Superficially, *cyber-human systems*, a new term from the U.S. National Science Foundation, doesn't sound closer to architectural robotics; however, a diagram capturing its essence also captures the essential components of architectural robotics: an expanded field of people, physical things, and computing strategies organized by a higher-level triangulation identified as the *computer*, the *human*, and the *environment*, the latter of which includes environments that are virtual, physical, and a hybrid of both.[3] While architectural robotics can identify itself easily in the sprawling field of cyber-human systems, so can all the other research communities mentioned on this page, which is surely—wisely—the point of redefining this core sponsored-research program that once bore the names human-centered computing (HCC) and human-computer interaction (HCI). Clearly, the cyber-human system (CHS) organization recognizes computing as expanding its reach and making more and new connections, which is good enough reason for those readers, aligning themselves with one of the research communities I just mentioned, rightfully to find my reductive characterizations lacking.

Another way to define architectural robotics is to list its basic ingredients, their qualities, and their individual contributions to the whole. By referencing "ingredients," we find ourselves in the realm of analogy; and there may be no better *analogical city* for collecting ingredients than Paris. Provided you can manage the 25-minute stroll, passing over the Seine and through *Île de la Cité* in course, the ingredients for architectural robotics can be found here: one part *Centre Pompidou*, one part *Fontaine des automates*, and one part *Bibliothèque Sainte-Geneviève*.

To architectural robotics, the *Centre Pompidou* (1977) brings the notion of architecture-as-scaffolding: a steel exoskeleton serving as the armature for the display screens enveloping it, the walls and floors reconfiguring within it, and the computing capacity permeating it—all of these being plug-n-play and easily or automatically updatable. Recognizing change as permanent and inevitable, the young architects Renzo Piano and Richard Rogers envisioned, in their architectural drawings submitted to the open design competition for the *Centre Pompidou*, a physically reconfiguring "communications machine," reprogrammable in both its physical and informational aspects. Their winning design was surely indebted to two unrealized architectural projects that preceded the *Centre Pompidou* competition by a few years: Cedric Price's Fun Palace (1961), a vision of an articulated, dynamic structure promoting interactions between it and its users that cause the physical form of the building to reconfigure[4]; and Archigram's Plug-in-City (1964), a mega-structural steel armature for receiving standardized plug-in components, such as living-unit pods.[5]

In a city government press release, the *Centre Pompidou* was offered, with great anticipation, as the physical manifestation of the hopes (and maybe the trepidations) of Europe in the mid-to-late 1960s: "A true science of information is now beginning to develop in correlation with the new orientation of science and the social sciences: art history, communications, cybernetics, linguistics and semiology have restated the concepts of theory, history, space and time, and of the symbol in new terms."[6] In this context, the *Centre Pompidou* was not to be "a passive cult-object" exemplified by the great works of art commissioned by noblemen, rulers, and religious leaders, but rather a grand public space open to "information, dialogues and debates."[7]

The architects' vision of the building-as-scaffold, sheathed in display screens and mechanized for physical reconfiguration, was never realized in the *Centre Pompidou* we know today. Nevertheless, the building has largely fulfilled President Pompidou's call for a "polyvalent cultural center" situated in the historic center.[8]

Undoubtedly, the building remains a daring presence in the historic city. But would the *Centre Pompidou* get built today? Andy Sedgwick of Arup, the engineering firm that played an essential role in the *Pompidou* design, doesn't think so.[9] "I don't think a national government would put its faith in such a young team today," opines Sedgwick. "In fact," he adds, "in our checklist-driven, bureaucratic world, I doubt [Piano and Rogers would] even make it through pre-qualification. With the industry now much more risk-averse than in those days—particularly public sector clients—many inspirational projects probably fall unnoticed at this first hurdle."[10]

While Sedgwick recognizes the obstacles in global terms, in Paris itself the pendulum has unmistakably shifted toward privatization and exclusivity as France endures an uneasy economic and political climate.[11] One obvious index of this shift is Renzo Piano's newest, grand addition to Paris: the *Fondation Jérôme Seydoux-Pathé*, bankrolled by Pathé and its parent company, Vivendi, the French multinational mass-media and telecommunications company headquartered in Paris.[12] The headquarters of this privately funded cultural foundation for Paris, a glass-covered, organic form tucked into the courtyard of nineteenth-century building fabric, cannot be described in the terms lavished on the *Centre Pompidou* by the design jury awarding Rogers the 2007 Pritzker Architecture Prize. The *Centre Pompidou*, according to a statement offered by the Pritzker jury, had "revolutionized museums, transforming what had once been elite monuments into popular places of social and cultural exchange, woven into the heart of the city."[13]

If the kind of late-1960s architectural ambition represented by the *Centre Pompidou*, the Fun Palace, and the Plug-in-City are precluded in the current social-economic-political milieu, local and global, it might yet occur at the much more modest physical scales elaborated in this book—as rooms, portions of rooms, suites of furniture—and as pavilions and small buildings. Along with material culture, generally, architecture has increasingly become a "marketing tool" for the private elite; perhaps the civic domain no longer has the stomach for such adventures at the grand scale of the urban, public monument anymore. This is what it is. *C'est la vie*. But maybe a rightful place for the vision, in lieu of today's "starchitecture," is where it can be of service as a kind of "chamber piece" to users conducting their everyday lives, at work, at home, at school, in the medical clinic, at leisure, and at play—an intimate marriage of architectural machinery and people, dispersed in space, rather than the insidious marriage of the masses and architecture as an expression of a corporate-cultural machinery, governed by the most powerful people and their private enterprises (and fantasies).

One key aspect of Piano and Rogers's competition entry that wasn't taken away in the *Centre Pompidou*, as it was constructed in Beaubourg, Paris, was the large expanse of building site left open by the architects as a public square, animated since its inauguration with street performers, carnivals, caricature and sketch artists, skateboarding competitions, and, most notably beginning in 1983, the *Fontaine des automates*. Known also as the Stravinsky Fountain, the *Fontaine des automates* features sixteen whimsical moving and water-spraying sculptures representing themes and works by composer Igor Stravinsky. The black-painted, kinetic sculptures mostly composed of repurposed machine parts are the work of Jean Tinguely and the colorful, organic forms the work of Tinguely's second wife, Niki de Saint-Phalle.

While the *Fontaine des automates* is highly photogenic (figure 12.1), a favorite photograph of mine from the extensive photo documentation of Tinguely's *oeuvre* is one (figure 12.2) capturing a mature lady, donning a fashionable outfit with matching hat, thoroughly enjoying her physical engagement with (i.e., pedaling) a black-painted machine-sculpture, much in the character of those found at the *Fontaine des automates*. Collectively known as *metamechanics*, these kinetic machine-sculptures of Tinguely count as my second "ingredient" of architectural robotics, demonstrating that machines, despite their weight, have a lighter side.

12.1
Architecture-as-scaffolding, accommodating display screens and moving parts. The *Centre Pompidou* with the *Fontaine des automates* in the foreground (photograph by the author).

12.2 (opposite)
The pleasure of machines. A visitor joyfully pedaling Jean Tinguely's *Cyclograveur* (1959) at the exhibition *Rörelse the konsten* ("Movement in Art") at *Moderna Museet*, Stockholm, May–September 1961 (photograph by Lennart Olson, reprinted with permission of the Halland Art Museum, Halmstad, Sweden).

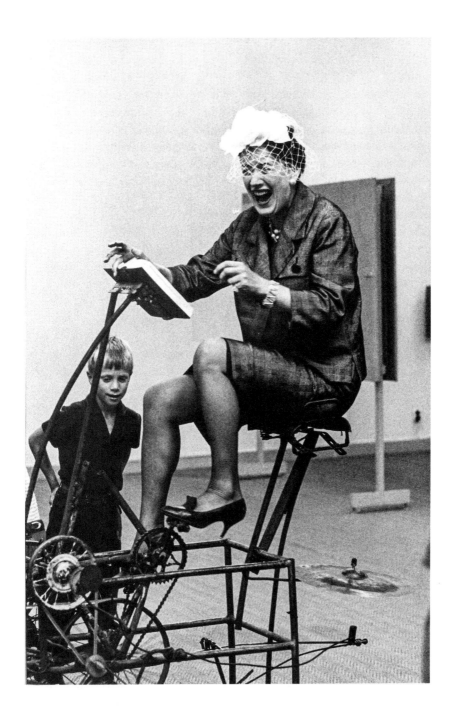

Ecosystems of Bits, Bytes, and Biology

Declaring all machines "art," Tinguely celebrated the decorative and sculptural aspects of machines as much as their functional capacities. Following this strange belief, Tinguely created machines that negate their characterization by industrialists and the lay public as utilitarian, dependable, and cost-efficient. Unlike the apparent grace and choreography of industrial robots, Tinguely's machines move about awkwardly, swaying, rotating, ringing, scratching, expanding, and collapsing. Most excitingly, they sometimes dismantle, deconstruct, and, by design, destroy themselves. Working hard but accomplishing nothing serviceable, these machines nevertheless offer us a lot in the way of delight and awe. In this meticulous amalgamation of gears, pulleys, belts, motors, and a host of found scraps scavenged from the mechanized world of old, we discover something of ourselves. (Giorgio de Chirico, the artist considered in chapter 6, insisted the same in his paintings of animated human figures assembled from scraps of material and mundane objects, mostly drafting instruments and broken picture frames.)

Such useless aspects of architecture and machines serve a very real function: to enchant us, to stimulate us, to provide us with a sense of wonder and, perhaps, even, a sense of hope. As architect Robert McCarter argues, "if we accept the machine as neutral, as something we can use 'without thinking,' then we risk being delivered over into a type of being completely determined by technological evaluations," where "meaning [is] replaced by mere means: utility."[14] Architectural robotics draws from Tinguely the notion that machines are very human and have something important to tell us.

The third ingredient of architectural robotics offered by Paris is Henri Labrouste's *Bibliothèque Sainte-Geneviève* (Paris, 1838–1850). From the *Centre Pompidou*, the *Bibliothèque Sainte-Geneviève* lies a straight-line south, running perpendicular to the Seine, with *Île de la Cité* marking the line's center so that the two buildings are geospatially a reflection of one another. This odd formal coincidence leads us to recognize one more common ground: the imposing, blocky façades of the *Bibliothèque Sainte-Geneviève* and the *Centre Pompidou* were designed alike as mechanisms for communicating with the public *by text* rather than by architectural conventions such as ornament, physical massing, and formal composition. Following here is a brief explanation of how this "display wall" façade comes about as an architectural icon of the *École des Beaux-Arts*, rather than as one in the shadows of the French civil unrest of May 1968, where it was intended but not realized.

Recall the opening words from chapter 10—"This will kill that"—by which Victor Hugo prophesied that architecture, as an expressive medium, would quickly become supplanted by printed media. According to Victor Hugo in *Notre-Dame de Paris*, printed media, produced mechanically by the printing press, could convey different points of view and changing perspectives endemic to the times; in contrast, an architectural work, being made of stone, offered a fixed means of expression that was hardly optimal for the times. Hugo had in fact asked Labrouste to comment on a draft of this work, which we know in English as *The Hunchback of Notre-Dame*, and, after his review of the manuscript, Labrouste expressed to Hugo his sympathy with Hugo's arguments. What Labrouste attempted to do in his next architectural design commission, for the *Bibliothèque Sainte-Geneviève*, was to make architecture once again speak. To renew the expressive promise of architecture, Labrouste transforms the building envelope, strangely, into the printed page of a book. Or, to be precise, in Labrouste's design, the façades of the library literally become the written index to the library's contents. Enveloping the building are the names of the major authors held in the collection, 810 altogether, inscribed in the stone façades at a scale visible from the street (figure 12.3), their letters originally painted red as if they were printed on paper. (The red type is an important quality that we miss today in *Sainte-Geneviève*'s façades.) Each author's name appears on the location of the exterior building envelope corresponding to the bookshelf within the reading room interior that contains that author's books.

12.3
Architecture with an added layer of information. Detail of the *Bibliothèque Sainte-Geneviève*—its card catalog inscribed in the façade (photograph by the author).

Labrouste transforms this building into a written guide to the library: stone is paper and binding, walls are shelves, light is illumination, columns are trees of knowledge, and the façades are a table of contents. *The building is a book. Words become worlds.* From the *Bibliothèque Sainte-Geneviève*, architectural robotics draws the communicative power of architecture—architecture of a different, less pure variety that relies (as did Labrouste's library) on it being hybridized with other information and communications technology.

SCENES FROM A MARRIAGE: A STRANGE HOMECOMING

So, what's in this marriage of architectural robotics and its inhabitants? As elaborated in this book, architectural robotics promises to accomplish two things personal computing hasn't done very well over the course of its short life. First, architectural robotics sets out to promote the social aspects of living: bringing together people, minds and bodies. Second, architectural robotics sets out to preserve the primacy of touch: using digital technology, curiously, to maintain human interactions with printed media and, overall, with physical things. These two capacities are contrary to the workings of the most dominant forms of human-computer interaction today, which has us tethered to smartphones, video-game consoles, MP3 players, laptop computers, and (less so yet alarmingly) to robotic companions during those periods of our lives when we are most fragile. As repeated in this book, such highly consumed forms of computing don't do much for the way we've conducted our lives for more than 200,000 years as social, tactile, and emotionally intelligent beings inhabiting a physical world.

This is not to say that architectural robotics is motivated by a dim view, by the kind of social breakdown that defines the literary genre—cyberpunk. Not everything is wrong with human-computer interaction. Quite the contrary, HCI has done much to connect demographically and geopolitically underserved populations to the global conversation and marketplace; HCI increasingly engages human touch, manipulation, and other sensory acts that are central to being human; and HCI seeks ways to lessen our impact on the environment by helping us manage and generate resources (where we can). Architectural robotics is not meant as a replacement or substitute mechanism for achieving these ends; instead, it is envisioned as one aspect of an ad hoc information and communications technology (ICT) system: a capacious cyber-physical home for a place-based form of computing, for human-machine interaction, and for human-machine intelligence.

An apt vehicle for defining this different kind of human-computer interaction, particularly given this chapter's repeated reference to printing, is Walter Benjamin's "The Work of Art in the Age of Mechanical Reproduction" (1936), an essay that exhibits a remarkable capacity for maintaining widespread cultural relevancy. Here, the key point for architectural robotics is that "Architecture ... is consummated by a collectivity in a state of distraction."[15] Architecture's physical scale—bigger than an object—allows many people to engage it. And in engaging it, we are caught up in it, in its spatial dimension and in its physicality. (That is, unless we are tourists, gazing at it from a distant point that effectively diminishes the architectural work, relative to our perspective, to an object.) If we are not so far from it, but rather in its shadow or inside it, an architectural work is large enough in scale to challenge (impose on) the physical scale of our bodies. Architecture surrounds us: it is spatial by definition. As such, we can't physically command architectural works, as we (believe) we can objects. Likewise, we cannot possess architectural works in the same way we can objects. (We can't travel with architectural works or display them on a shelf.) This means that unlike an object—Benjamin uses the example of a painting—an architectural work, given its physical scale and its spatial dimension, affords a "simultaneous collective experience."[16] Architecture accommodates the individual, yes, but also the group. It is characteristically a social space for both minds and bodies—*for us*. Two-dimensional, screen-based artifacts and cyberspace cannot fully and adequately replicate this quality of space.

In this essay, Benjamin emphasized that the interaction between architectural works and their inhabitants is both visual and tactile.[17] We recognized earlier in this book that the interaction between architectural works and their inhabitants is not only visual and tactile but also auditory and olfactory. Blindfolded and still, we can reasonably guess the gross dimensions of a room by listening for the reverberations the enclosure makes, and we can sometimes, by smell, distinguish the primary construction material of a building, be it concrete or wood (or, if neither of these, we might conclude that it's made of glass and steel). Different from our interactions with object-things that are fundamentally visual, "this mode" of interaction, between an architectural work and its inhabitants, wrote Benjamin, "in certain circumstances acquires canonical value. For the tasks which face the human apparatus of perception at the turning points of history cannot be solved by optical means, that is, by contemplation, alone. They are mastered gradually by habit, under the guidance of tactile appropriation."[18]

It is not a simple task to unpack Benjamin's purpose in this fragment, but it suggests, first, that an architectural work is something we come to know by habit, by feel, by living with it, rather than by "contemplation alone" (and so, architecture is "*habit*-ation"); and, second, that our understanding of unfamiliar phenomena (e.g., a new technology) can only be accomplished in the same way that we come to know an architectural work: by feeling it, by getting a sense of it over time, by letting it grow on us, by making it a habit, by inhabiting it.

Let's stay here for a moment with this concept of habit and habitation and consider a simple case: a wool cap. A wool cap doesn't mean much by itself on the sales shelf or sitting in your shopping bag immediately after its purchase. But on a cold day, once you handle it with your hands and perceive the warmth it will provide, it begins to mean something to you. And you really come to appreciate that wool cap on a cold day, when it's on your head, helping to maintain your warmth. You *grow* to appreciate it. It *grows* on you. You value it over time, by use, by a habit largely based on "tactile appropriation." (Indeed, if you're wearing a wool cap on your head, you can't see it.)

If you exchange the wool cap for a smartphone, then habit plays arguably a larger role. Even if the smartphone has a striking or elegant visual design, you don't set a smartphone on the table next to you only to admire its sparkle. Nor do you simply place a smartphone in your pocket as you put a wool cap on your head. Instead, you come to know the many features and affordances of the smartphone over time by use of your hands and fingers: the weight of the phone, the temperature and texture of the phone's materiality, the extent of its screen, the position of icons on the screen, the mechanical toggle switch for audio, the mechanical push-button for power, the location of the mini-jack for your headphone or your headset. So while the striking visual design of a smartphone is integral to your desiring it, it is only through habitual use that you *grow* to appreciate it; that it *grows* on you. And despite its nearly countless affordances, you and your smartphone develop, over time, a particular relationship: patterns of use that suit you both. It knows quite a bit about you, too. You work together.

In a compelling consideration of Benjamin's essay and its implications for architecture, architectural theorist Stan Allen describes the interaction between an architectural work and its inhabitants as an "unconscious intimacy."[19] It's a matter of how it feels, how it makes you feel, the sense of it. From this unconscious intimacy, from this habitual use, argues Allen, we become informed users of an architectural work, "adept in reading … its clues to behavior."[20] If we replace "architectural work" with "digital technology" to capture something of the *robotics* in architectural robotics, it becomes this: from our unconscious intimacy with digital technology, from habitual use, we tech-users become "adept in reading … its clues to behavior."[21]

With architectural robotics, its inhabitants become adept in discerning a rather complex, cyber-physical environment that is, at once, something familiar (the architectural character of it; and maybe the nuanced, life-like movements of its compliant forms) and something not (its curious hybridization of architecture and robotics that makes it cyber-physical). Indeed, toward supporting and augmenting our increasingly digital society, a design strategy that seems most apt, to borrow words from Stan Allen, is one that offers "something unrecognizable out of the familiar."[22]

From the outset, this book has shown that it is not only architectural robotics that is both familiar and strange to us, but arguably *all everyday artifacts*: the furniture at that seaside hotel that seemingly comes to life, the ghostly row house that stands vacant, the walls and ceilings of your home overtaken by disturbing shadows after your weekend away. Alfred Hitchcock made a career of manipulating this palette of domestic architecture and its many appearances, which included, as well, capturing in film the built environment from awkward angles that we rarely experience in our everyday habits. These unsettling experiences rely on presenting to us the most unfamiliar dimensions of familiar places. These unsettling experiences unsettle us (or alternatively, amuse us, as we considered earlier in this book in chapter 6) precisely because they challenge our notion of control. How can we be so frightened or so amused by an environment so familiar to us? Our own home?

Like Alfred Hitchcock for artistic purposes, architectural robotics harnesses this strangeness for productive ends. Today more than ever, the unfamiliar aspects of a technology-embedded environment suit us. Our increasingly digital world has prompted an existential awareness: that intelligence does not reside solely within us, but emerges partly from our interactions with a built environment that, inevitably, will become increasing computational, physically reconfigurable, and intrinsically more a part of us.

Fittingly, N. Katherine Hayles describes my *home+* as a "distributed cognitive environment."[23] As defined by Hayles, a distributed cognitive environment is a technologically embedded environment capable of "extending embodied awareness in highly specific, local, and material ways that would be impossible without electronic prosthesis."[24]

In defining a distributed cognitive environment as such, Hayles refers to an interpretation of John Searle's famous "Chinese room," offered by Edwin Hutchins, professor of cognitive science at the University of California, San Diego.[25] In Searle's thought experiment, the Chinese room is occupied by a person (Searle) who has no knowledge of Chinese. In a kind of Turing test, questions written in Chinese are slid under the room's door, and Searle is asked to respond to these inquires in Chinese, using a basket of written Chinese characters and a rulebook for deciphering and assembling Chinese texts.[26] Using only these material resources, Searle—remember, he doesn't know Chinese—is nevertheless capable of producing clever responses to the Chinese inquiries, suggesting to those outside the room that the Chinese room's occupant understands fully the conversation, when in fact he has no understanding of the written exchange unfolding. Employing the Chinese room as a metaphor for computing, Searle argues that we cannot program a computer to exercise human understanding or consciousness (which is the so-called strong AI position); we can only simulate these. But can't simulating understanding and consciousness in a machine do enough for us? Does the holy grail of computing need to be the achievement of conscious *understanding* in a machine? (To what ends? What are the ramifications of achieving this?) Can't we respect that human cognition is anyhow both conscious and unconscious, concrete and abstract, intuitive and conceptual? And if cognition is all of these things, doesn't a machine like architectural robotics exhibit something of human cognition, given that human cognition is partly unconscious?

The significant point for Hutchins is not that machines can't be programmed for human understanding (Searle's argument) or that the human being occupying the room proved to be clever enough to be convincing, but rather that *the human being and the room, working together*, proved to be so clever. According to Hutchins's theory of "distributed cognition," cognitive activity doesn't reside solely in an individual but instead is distributed across other people and things surrounding that individual, as well as the "the environment," the latter characterized as a "material" (physical) environment that is, itself, "a computational medium."[27] Similarly, but from a robotics perspective, Hans Moravec acknowledges the role of the physical world in the construction of meaning, postulating that "if we could graft a robot to a reasoning program, we wouldn't need a person to provide the meaning anymore: it would come from the physical world."[28] For both Moravec and Hutchins, the physical environment is not something outside human consciousness but rather an integral part of it and a primary source of meaning.

What for Hutchins is a *cognitive room* then becomes for N. Katherine Hayles a distributed cognitive environment.[29] In a distributed cognitive environment—my *home+* project, remember, is a fitting example for Hayles—"humans and intelligent machines ... interact in hundreds of ways daily," forming a "dynamic partnership" in which "'thinking' is done by both human and nonhuman actors."[30] In ascribing thinking to human and nonhuman agents working collaboratively, Hayles is not suggesting the gloomy kind of post-humanism that makes for visually engrossing science fiction movies populated by frightfully mindless cyborgs, but rather something affirmative for an increasingly digital society. "The prospect of humans working in partnership with intelligent machines," maintains Hayles, "is not so much a usurpation of human right and responsibility as it is a further development in the construction of distributed cognition environments, a construction that has been ongoing for thousands of years."[31] For thousands of years, indeed, the human-machine partnership has been dedicated to "freeing humans to give their attention to other matters."[32] This kind of freedom is especially coveted today, given that human attention, amid all of our digital connectivity is, as Hayles knows, a "scarce commodity."[33]

Hayles draws a more comprehensive model of cognition, defined by a pyramid in which "human consciousness would ride on top of a highly articulated and complex computational ecology in which many decisions, invisible to human attention, would be made by [and with] intelligent machines."[34] Hayles affirms that "to conceptualize the human in these terms is not to imperil human survival but is precisely to enhance it, for the more we understand the flexible, adaptive structures that coordinate our environments and the metaphors that we ourselves are, the better we can fashion images of ourselves that accurately reflect the complex interplays that ultimately make the entire world one system."[35] Here, Hayles, too, reminds us that we are fundamentally superadaptors to both environmental and psychological conditions.

Notably, meaning in Hayles' model of cognition is derived not only from the top of the pyramid (where human awareness resides) but also from "lower-order" cognition that constitutes both the bottom of the pyramid and its greater area. Following from this conception of meaning offered by Hayles, we can conclude that the interactions across people and their surroundings cultivated by architectural robotics contribute meaningfully to the construction of places meaningful to us. In this canted light, architectural robotics represents a vague yet affirmative project: something of the old, place-based architecture, combined with newer information and communications technology that, collectively, serve as a capacious, distributed, "cognitive" home for people and things of all kinds.
From here, we consider the future course of architectural robotics, beginning (again) with what *isn't* the future course of architectural robotics. The future of architectural robotics isn't situated in the World Wide Web, or in the smart city, or in *e-topia*, or in cyber city. Instead, we might say—in a bit of a silly way—that the future of architectural robotics occupies *cyberPLAYce*. The vision of a playful architectural robotics moves the Internet toward the *intra*net, but hardly as far as Thoreau's cabin. The future of architectural robotics is not yet, altogether, disconnected from the Internet. (It anyhow may not be possible to disconnect from the Internet, and, arguably, it's not entirely desirable.)

More precisely, the future of architectural robotics will rely less on the *Net* and more on a cyber-physical *mesh network*, a communications network far looser than the Internet: less formalized, less oligarchical, more decentralized, more community-responsive, more ad hoc. Like a wireless mesh network, the more expansive (and playful) plan for architectural robotics recognizes the architectural robotic environment as a node of activity that may or may not form a vague connectivity with other nodes to form an *architectural robotics mesh network* (ARM-N). In defining an architectural robotics mesh network, let's refer to an architectural robotic environment just under way in my lab that would especially benefit from such a network: *I-cubed*, a physically reconfigurable, technology-rich, mobile, architectural robotic library, 10 feet on a side, serving (in particular) underserved communities. Accordingly, a single I-cubed, a single architectural robotic node, can exhibit a variety of connectivity conditions (figure 12.4):

- It may function off-line.
- It may join the Internet for increments of time or continually.
- It may form a mesh network—for increments of time or continually—with other I-cubed nodes or, maybe, I-cubed nodes and an AWE at the library, staffed by librarians).
- It may form a mesh network (as above, for instance), which in turn connects to other mesh networks for increments of time or continually.

Akin to the reconfigurability exhibited within one architectural robotics environment (one node), the larger-scaled architectural robotic system, a mesh network of architectural robotic nodes, can form and restructure itself and can also form and restructure its connectivity to other mesh networks as well as to the Internet.

In an architectural robotics future, the information transmitted across architectural robotic nodes doesn't flow evenly throughout the mesh network, but "hops about" to neighboring nodes, as affinities arise from interactions between itself, people, and the things around and inside it: that amalgamation of physical bits, bytes, and biology at a given instant. In this way, architectural robotics creates a small home—*your home, your neighborhood*—for computing, which makes architectural robotics the scene of a strange kind of homecoming.

What might the plan of this architectural robotics mesh network look like? In concept, something like the mesh network formed by localized XO laptops, the vision of the same Nicholas Negroponte who looms largely in this book. In a screen shot of such a mesh network, taken from an XO laptop (figure 12.5), each node—a child's computing universe, represented by a small cartoon figure of her or his color choice—becomes the child's home to computing resources, locally and globally. If we recall the characterization of space by Deleuze and Guattari as *striated and closed* or *smooth and fluid*, then this small universe of connecting, computing nodes exemplifies the latter. To clarify by way of analogy again, a mesh network of XO laptops or architectural robotics is like a field of benday dots that, on close inspection, looks hazy and formless but, from a distance, forms a distinct image.[36] Compare the field of tiny benday dots spread over a printed page in a comic book (or, alternatively, on a canvas painted by Roy Lichtenstein) to the unequivocal meaning of that enormous, uninterrupted

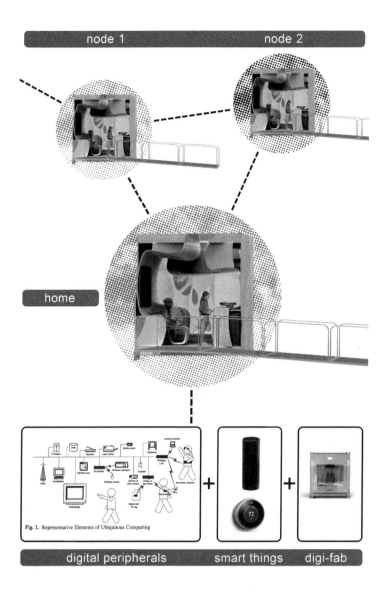

12.4

An example of an architectural robotics ecosystem. Shown within the three circles (one larger and two smaller) are three I-cubed nodes—each containing other digital devices, and all networked to each other and to still other I-cubeds and other kinds of architectural robotic environments not shown but infered by the extended black lines (diagram by the author).

circle that Apple Inc. built for itself in Cupertino, California, however elegant and transparent its glass skin. The medium *is* the message.

Similarly, compare the scene of figure 12.5—our ad hoc village of small homes to computing displayed on an XO laptop—to that of figure 1.2 showing an array of interconnected devices that defines ubiquitous computing. Interconnected digital devices like those of figure 1.2, combined with an expanding population of nominally intelligent, "smart" devices, represent the hardware aspect of *cloud computing* on a global scale.[37] While these Internet-connected devices are inside your home, workplace, school, or hospital, their control comes primarily from outside. There isn't (at least, not yet) a user-controlled "privacy setting" for them (other than cutting their cables). For now at least, such devices may be perceived as benign annoyances, intent on selling you things (as does, say, Amazon's *Echo*) under the guise of being *of service*. But the future is expected to bring you many more, much smarter, and potentially more invasive smart devices connected to the Internet.

What's occurring now on the Internet, as reported by the *New York Times*, the *Washington Post*, and (unashamedly) the oligarchies of the Internet themselves, is a frightful indication of what will inevitably encroach on the "Internet of Things."[38] A sample from the *New York Times* report:

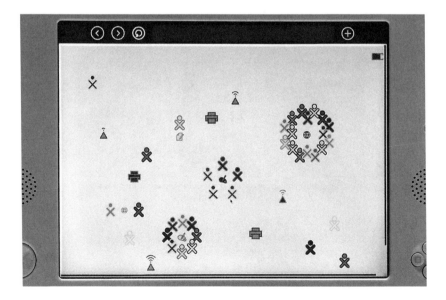

12.5

A mesh network displayed on the screen of an XO laptop (source: Wikimedia Commons).

- Uber infers that one of its customers had a one-night stand (it calls them "rides of glory") when a car registered with its app delivers him or her, one evening, to an address other than his or her home, and collects him or her, the next morning, at the same address. Uber alleges to be using this kind of information against journalists critical of its business practices.
- OKCupid, by its own admission, occasionally matches users with an incompatible date to help it learn what will happen to the pair on its dating site.
- Target sends its customers coupons for baby gear when its sales data suggest they're pregnant.
- Facebook (which admits to experimenting "on every user at some point") adds an "I voted" button that increases voter turnout in a congressional election by 340,000 votes, a figure large enough to decide an election at any scale.

These kinds of sites and apps "make it their business to track what we do, whom we know and what our typical behaviors and preferences are," which can impact our employment prospects, our credit reports, our intimate lives, and our democratic institutions.[39] What "we gain from digital connectivity," the *New York Times* report concludes, we lose in "the accompanying invasion into our private lives which makes personal data ripe for abuse."

This invasion by the Internet, and its inevitable expansion to hardware, has sounded the alarm of some very bright people, like Stephen Hawking and Elon Musk. Hawking and Musk, who speak on behalf of many other science and technology leaders, fear that the exponential growth and reach of machine intelligence, combined with its lack of morality, may position AI to rule and even extinguish us.[40] While it may be no more than a modest recourse, the envisioned plan for an architectural robotics mesh network establishes scale relationships for computing (see figure 12.4), much as Vitruvius, some two thousand years ago, established for architecture. Working from the bottom rung of this scale upward, we begin with the single device (like those shown in figure 1.2 for ubiquitous computing), which is then collected together in multiples under the "roof" of an architectural robotics node (e.g., one *home+* environment), itself collected together in multiples as a mesh network "neighborhood" (e.g., one hospital wing composed of multiple *home+* nodes), themselves collected together again as a metropolis of mesh networks—each rung of the ladder having more or less connectivity across it and to the other rungs, as well as to the Internet, as circumstances demand. With further advances in biological and quantum computing, nanotechnology and programmable matter, this ladder would extend to lower rungs yet, of still smaller physical scales. As a result, the communications at-scale and between scales may become more efficient and awe inspiring.

In architectural terms, this informal architectural robotics mesh network exemplifies a *weak architecture*, a theoretical concept elaborated by the late architectural theorist Ignasi de Solà-Morales. In a *weak architecture*, an architectural work and/or its components "flutter to and fro, like a spirit, promoting a kind of accord'" with other architectural works elsewhere.[41] Needless to say, by characterizing an architectural work (architectural robotics or not) as *weak* in no way suggests that it isn't thoughtfully designed or that it is insubstantial or insignificant; rather, such an architectural work is a meticulous, artful, if somewhat indefinite design, sharpened by the interaction, the performance, of human inhabitants and the movement of things in and around it.[42]

To be clear on this point: designers of architectural robotics, whether at the scale of a node or a mesh network, are accomplishing two design feats: they are giving form to the interactions afforded by the architectural work, and they are attending fully to the design of the artifact itself. Interaction design and aesthetic design, I argue, are not separable; and my case is made by citing Apple's decision to entrust product designer Jony Ive with the design of its mobile software platform, iOS 8, a charge previously led by Apple's software engineers.[43]

A promising design strategy for realizing an architectural robotics mesh network was defined as "urban acupuncture" by Solà-Morales.[44] In "East of Planning: Urbanistic Strategies Today," Robert Kaltenbrunner elaborates this design strategy that may prove vital to realizing an architectural robotics future at a larger scale. This strategy, writes Kaltenbrunner,

> ... must first and foremost grasp urban space as a differentiated whole, in which not everything has to be everywhere at once. [...] Such a strategy is comparable to acupuncture. It is localized, concrete and applied on a case-by-case basis, yet it always has the overarching context in view and attempts to guide its progress even if indirectly and over the long haul. A surgical intervention here, an autonomous implantation there—the effects radiate outward to the surroundings, gradually linking up with them and eventually forming a consistent urban structure.[45]

Borrowing a term from Michel Foucault, Kaltenbrunner and allied architectural, urban, and human-geography theorists describe this kind of weak organization as a "heterotope," a structure characterized by some numbers of seemingly vague yet recognizable and meaningful relationships between its constituent parts. An example is found in Barcelona, in its network of more modest architectural works contributing (quietly) to the city's redevelopment efforts in the years since its preparation for the Olympics. As applied within Barcelona and also far from it, "urban acupuncture" is intended to produce "small scale but socially catalytic interventions in the urban fabric."[46] This design strategy represents not weakness but, for Erik Swyngedouw, an "active, democratic, and empowering creation" that "returns the city and the city's environment to its citizens."[47]

For architecture as much as for computing, the critical concerns for the near future are *how*, and *to what extent* will its different scales dynamically interconnect, which can be defined, in one word, as *scalability*. Imre Horváth, professor of design engineering at the Technical University of Delft (TU Delft), defines scalability and the many forms it assumes in this way:

> **Scalability may be contraction (down-scaling) or expansion (up-scaling). [...]** *Geographic scalability* **involves maintaining performance, usefulness, or usability regardless of expansion from concentration in a local area to a more distributed geographic pattern.** *Loading scalability* **means expanding and contracting the resource pool to accommodate heavier or lighter loads or inputs.** *Administrative scalability* **concerns increasing the number of users or organizations to easily share a single distributed system.** *Instrumental scalability* **is enhancing the ease with which a system or component can be modified, added or removed.**[48]

What Horváth makes clear is the reach and significance of scalability in our networked world. Fittingly, Horváth repeats in the same paper the critical concern I posed at the head of this paragraph, albeit in more technical terms: that "the primary design question is how to architect a complex system to be extendable to multiple arbitrary scales in time and space."[49]

For architectural robotics, the way it might work is this: new algorithms will scale the offerings of the Internet to localized, distributed ecosystems, digitally grounded by architectural robotics. This digital grounding is of a different, more intense flavor than that adroitly elaborated by Malcolm McCullough, which considers computing in the built environment largely *as it is* without the catalysis of a novel kind of architectural artifact such as architectural robotics.[50] In an architectural robotics future, the computing resources delivered to its localized, distributed ecosystems by the Internet will "decouple" and "re-engage" to relative degrees, as determined by local situations defined partly by human-machine interaction and partly by individual or community control. How devices receive, transmit, and interpret signals from one another within and across nodes is defined as "interoperability on a systematic level" by Paul Kominers in a paper for Harvard's Beckman Center for Internet and Society.[51] The interoperability across larger-scaled architectural robotics networks and the Internet has the potential to provide more apt and hopefully more assistive computing resources to individuals and their local communities. What the service providers receive in return for relinquishing a bit of their pervasive control to local mesh networks is focused, localized tutoring of their AI platforms, as well as the knowledge required to deliver more focused content to local "neighborhoods."

Something of this friendly tug-of-war occurs in the energy industry, now that individual homes, multifamily residential buildings, and neighborhoods can harvest energy to fulfill their needs through localized solar collection; purchase energy from large oligarchical utilities when self-generation falls short; and sell energy to the same oligarchies when self-generated energy exceeds local energy needs. It seems reasonable that something of this model can extend to computing resources, recognizing, however, that energy provisions work in the background of our lives, and that the structuring of computing resources will be more complicated, given their foregrounding in our everyday lives.

In a scalable context like this, digital artifacts of all scales and capabilities, collectively, can create a commodious home to your local computing culture, to locating your meaning within a broader networked world while maintaining some locus of control. In short, architectural robotics domesticates computing, while the overwhelming tendency in computing is to globalize. An architectural robotics future promises a robust, dynamic, strategically decentralized, ecological, bottom-up platform for the cooperative interaction across computing, the built environment, and people. While architectural robotics, taken in this light, counts allies in hacker culture, DIY, and the maker movement, its future is more aptly characterized, given its complexity, by Imre Horváth's particular conception of cyber-physical systems. For Horváth, cyber-physical systems are "structurally and functionally open, context-sensitive, intelligent and self-managing engineered systems in which the physical and the cyber constituents evolve cooperatively ... with the social world [and] the [psychological] world of humans."[52] Horváth further elaborates the terms of his definition of cyber-physical systems as follows:

> *Structural openness* means that [the cyber-physical systems] may include collaborative subsystems of varying spatial scales and complexity scales—both in time and in space. *Functional openness* implies that they may consist of units that happen to enter or leave the collective at any time. The units: (1.) can be highly heterogeneous (computers, agents, devices, humans, networks, etc.), (2.) may operate at different temporal and spatial scales, and (3.) may have different (potentially conflicting) objectives and goals.[53]

What architectural robotics offers cyber-physical systems (CPSs) is some modest grounding, the fleshiness of architecture, as well as the long-standing significance of architecture as the sheltering and cultural abode to humankind.

As a hybrid of architecture and computing that provides that stage for interactions across people and things of all kinds, architectural robotics may be identified by a third, emerging current of CPSs that is not defined as an industrial system or a social system but rather, for Horváth, an "Experience System."[54]

A MARRIAGE OF ANOTHER SORT: "TECHNOLOGY AND THE HUMANITIES"
Let's return for a moment to the second of two epigraphs that appear at the front of this book, offered by a visibly frail Steve Jobs on-stage in San Francisco when introducing the iPad 2 to his devoted audience. "We believe it's technology married with the humanities," avows Jobs, *"that yields us the results that make the heart sing."*[55]

At this late point in these pages, it is constructive to recollect what of "the humanities" has been brought to this consideration of architectural robotics. Principally, the humanities is, in this book, represented by the vast cultural dimensions of architecture, beginning with the writings of Vitruvius from Roman antiquity and including the countless architectural artifacts that have accumulated on our planet before and since. In these pages, the humanities also materializes prominently as audacious thinking, across wide-ranging disciplines, accomplished by creative human beings who insist that they can envision and realize artifacts responsive to a dynamic and diverse society. Drawn from the humanities, the richness of architectural thought and production, combined with expansive, enlightened thinking on the workings of human beings in the worlds they inhabit, and how to accommodate and advance these, has served to position architectural robotics as a capacious stage for the theater of our increasingly wired, connected lives. There is, lamentably, a tendency in recent architectural production to forget architecture's riches, its purpose, and even the creative genius and high-level craft required to propel it. Architecture of late is too often preoccupied with form and its representation in the media, consumed as photographs under brand names like *sustainable* and *green* (to convince us that the architectural work is built with the best intentions) or *artistically motivated* (to impress upon us that the building represents a work of art that, in all likelihood, is financed by a wealthy, private patron.[56]

In other parts, architecture has begun operating something akin to industrial design in its employment of evidence-based design methods and post-occupancy evaluations, counterparts (on the "lighter side"?) to the methods of ethnography, human factors psychology, and usability engineering employed in the development of this book's three case studies, and commonplace to the activities of industrial design, interaction design, and HCI.[57] For architectural robotics, as for any human-centered design and engineering activity, there is the question of how much confidence one should place in user studies, and how much of design activity is otherwise the work of the creative, meticulous genius of architecture and engineering. Does a reliance on user input and codesign permit the Brunelleschi and da Vinci of today? Or is this prospect also precluded, as with the likes of the *Centre Pompidou*?

Any response offered by the technology sector (and by this, I include myself and my peers) might heed the warnings of Lewis Mumford, a mostly forgotten voice from the middle decades of the past century, in his condemnation of "pragmatic liberalism" that persists in cultural and political life still today—surely, in the United States. "This pragmatic liberalism," Mumford writes in "The Corruption of Liberalism,"

> was vastly preoccupied with the machinery of life. It was characteristic of this creed to overemphasize the part played by political and mechanical invention, by abstract thought and practical contrivance. And, accordingly, it minimized the role of instinct, tradition, history; it was unaware of the dark forces of the unconscious; it was suspicious of either the capricious or the incalculable, for the only universe it could rule was a measured one, and the only type of human character it could understand was the utilitarian one.[58]

Mumford's words illuminate many of the broader themes of this book: that designers of caring, intelligent, and appropriate human-machine systems must recognize the import of history, tradition and counterculture, of perception and consciousness (of all kinds), of exactitude, capriciousness and the immeasurable, of dreams and sorrow—in short, the stuff of life. In this respect, maybe, architectural robotics does have something in common with cyberpunk: its deployment of technology in unanticipated ways, in sometimes impractical ways, in ways that may be curious or fanciful, as well as in ways not precisely projected by its designers (Tinguely's metamechanics). In the late 1960s, there was at least the suggestion, the hope that technology might be a vehicle for freedom, not control; that technology was an invitation to shape and reconfigure one's life. Indeed, in the first issue of the *Whole Earth Catalog* (Fall 1968), Stewart Brand made this hope the magazine's "Purpose," offering that a "realm of intimate, personal power is developing—power of the individual to ... shape his own environment."[59]

In "The Corruption of Liberalism," Mumford lamented the passing of a previous era in which "there was as large a field for imaginative design and rational discipline in the building of a personality as in the building of a skyscraper."[60] In these few, disparaging yet astonishing words, Mumford captured a key theme threading through this book: the association of constructing a human personality, a human being, with constructing the architectural work, constituted by a process that is creative and scientific. If design researchers can bring their capacities for what Mumford calls "imaginative design and rational discipline" to the design of computational artifacts, then these same artifacts should, in turn, serve as vehicles for cultivating imagination and reasoning *in their users*.

This modest tactic for building a better world runs contrary to the tyranny of the intelligent "recommender systems" under the banner of artificial intelligence. Do we want computer intelligence, everywhere, suggesting what we might like on the basis of our previous choices? Are recommender systems advancing society or otherwise peddling the products of some very large, online merchants? Some years ago, touring Microsoft's House of the Future, which offered intelligent recommendations for nearly every home ritual (preparing meals, listening to music, selecting a bedtime book to read to your child, etc.), I entered the nightmare of being prisoner to my own preferences, in my own home. A popular cartoon from a 1982 issue of the *New Yorker* captures the same: it pictures a prisoner, strapped to a chair in a bare cell, forced to endure endless repetitions of Pachelbel's *Canon in D*, one of the most recognizable and popular classical music works. The familiar is nice, sometimes, but often stilting. It certainly doesn't cultivate imagination and reasoning in the way that Mumford encouraged. One has to wonder, then, whether Google, Facebook, Twitter, Microsoft, Amazon, Apple, and Alibaba are building us skyscrapers or prisons?[61]

And what of the physical materiality and design dimensions of the House of the Future? The interior looked to me extraordinarily ordinary, suggesting again that while artificial intelligence may be encroaching on human intelligence, this intelligence has no body, no flesh. Is computer intelligence returning us to the concept of a mind-body dualism? Do we want, as Descartes imagined in 1637, in his *Discourse on the Method*, to exist as not more than a mind occupying a small black box, where inputs and outputs can be controlled by some source outside it?

> **I decided to pretend that everything that had ever entered my mind was no more true than the illusions of my dreams, because all the mental states we are in while awake can also occur while we sleep and dream, without having any truth in them. [...] This taught me that I was a substance whose whole essence or nature is simply to think, and which doesn't need any place, or depend on any material thing, in order to exist.[62]**

Updating Descartes' argument for the information age, philosopher Hilary Putnam argues that if we spent our lives lying on a table with our brains receiving inputs sourced from a computer, consisting of a life lived in the world, we would not be able to distinguish the deception from the reality.[63] Indeed, we inhabit a world that is decreasingly physical and increasingly digital, so that "soon," as Harry Mallgrave concludes, "we may cease to desire to interact with the physical world."[64] In our increasingly digital world, the AI of a few oligarchical companies is increasingly providing us with the map of the physical world, such that "everybody follows the prompts and chooses to be like each other"; or alternatively, everybody chooses the recommendation that reaffirms who *they were* rather than to become *what they could be*.[65] In both cases, it means, simply, that "everybody" is engaged in "things that are easy and mentally undemanding."[66]

From so many disciplinary perspectives, this book has disowned the growing dependency on the "virtual map of the physical world," making clear that a brain with no body has limited access to the world, recognizing that *we are* superadaptors by nature. *A rebours*, what architectural robotics promises to do is to make a home for us in both *the virtual* and *the physical worlds*, to both expand and make accommodating the threshold between these realms, and to make passages from one to the other more seamless. One might argue that oligarchies such as Google, Amazon, and Microsoft are in the same business, reaching outside the screen and into the physical world, more and more, with new products and services. But architectural robotics, harnessing the capacity of architecture, can create its own place: a localized platform connected, whenever desirable or needed.

In introducing architectural robotics—an emerging subfield at the interface of environmental design and computing—this book, contrary to the pragmatic liberalism denounced by Mumford, has sought to make the necessary space between its covers for the coexistence of not only the virtual and the physical, the local and the global, but also the humanities and technology. As Steve Jobs knew well, user studies and design creativity lead nowhere without a vision of what's coming next, as well as a firm understanding of from where we came. Architectural robotics is dependent on the design-researcher, part prophet, part visionary, steeped in history and theory, art and culture, science and engineering, along with practical know-how and street smarts. Which is another way of expressing what Vitruvius did, some two thousand years ago, in the most concrete terms:

> **Architects who without culture aim at manual skill cannot gain a prestige corresponding to their labours, while those who trust to theory and literature obviously follow a shade and not reality. But those who have mastered both soon acquire influence and attain their purpose.[67]**

THE BEAUTY OF THE MACHINE

The vitality of architectural robotics—reconfigured, distributed, or transfigured—is not dependent on the forms and behaviors of the natural world, nor does it evolve linearly from an architectural precedent of hinged surfaces and sliding walls. Instead, architectural robotics, in its form and intelligence, follows its own logic.

For one, neither architectural robotic environments nor their components bear faces that look like our faces: eyes, nose, mouth, and ears, organized more or less symmetrically. As a consequence of not having a face, architectural robotic environments don't purport to be particularly intelligent. We don't look into their eyes and expect to be received as if by an intimate friend. In vague ways, however, architectural robotic environments may look, move, or otherwise behave like us or like other living things familiar to us, but overall they exhibit an artificial nature: their own in look, behavior, and intelligence—apt for what they are, what they're meant to be, and for their purpose. Architectural robotics is not a pretty face, but *artifice*—"life-as-it-could-be," to borrow the words of Christopher Langton.[68]

With architectural robotics, it's anyhow not much of an a-life (artificial life) and more of a low life. Recalling chapter 7 and Jeff Hawkins's consideration of intelligence, architectural robotics doesn't require an extraordinary degree of intelligence and certainly doesn't have to look intelligent. And this, being an assistive technology, may console someone like Sherry Turkle, who rightfully (to me, at least) becomes uneasy while witnessing an elderly, frail person offering intimate thoughts to a furry, robotic friend like Paro, the "therapeutic seal."[69] The ethics of Paro and its intelligent relations ("self-replicating nanobots," "armies of semi-intelligent robots," "autonomous weapons," and other breeds of intelligent machines that are expected to increasingly populate our planet) is a topic extensively considered in the lay press and adroitly considered by Illah Reza Nourbakhsh in *Robot Futures*, and (as such) is considered only narrowly here in this book and its closing chapter.[70] (The topic of educating students in architectural robotics, likewise, will not be considered in this book; my colleagues and I have published a paper in *IEEE Robotics and Automation* on this subject, available online and in print.[71]) The proposition to be made here, rather, is that the embedding of robotics in the stuff that's already part of our lives—furniture, rooms—is not likely to distance us from each other in the way a humanoid robot might.

And architectural robotics, without face or animal fur, may yet prove endearing to its users. Cars do. Homes do. And hasn't Apple made a fortune selling products doing so? Did we really buy that iPod or iPad because it was so much better performing than all MP3 players or tablets at a fraction of the cost? Or was it that we like the way it looked, felt, behaved in our hands—the seamlessness between its design and technology, "technology married with the humanities"? Here, architectural robotics benefits partly from our innate tendency to ascribe agency to inanimate things, to personify them. It is the inverse of John Hale's thesis in the *Future of the Future*: we don't only externalize what we are into the world but recognize ourselves in things of the world.

Without face or fur, with no or few words, behaving in a manner that is only vaguely familiar to us, architectural robotic environments and their components can yet endear themselves to their inhabitants. We already considered the central role habit plays in this, but what about the visual component of this relationship? What, in the appearance of architectural robotics, engages us, attracts us? An apt response comes from the Italian designer Alessandro Mendini and his *Stelline* chair.[72] Whereas the reference in Mendini's *Sirfo* table (1986) is plainly a goose supporting a tempered glass disk, his *Stelline* chair (1987) is not explicitly suggestive of the form of a person. Instead, the *Stelline* chair, in its few tubular elements, its few marks in space, suggests a human body in much the way a child draws a stick figure; only this being a chair, its figurative suggestion is just a little bit less obvious (figure 12.6). If a light source, from above and at the right angle, casts a shadow from the chair across the floor or wall, the user might discover there a cartooned "friend" drawn by the absence of light. In a sense, Mendini has designed the *Stelline* chair so that its users are, accordingly, designers themselves: designers in interpreting a familiar "code," the stick figure, that suggests to users the form of a person, foreshortened and in shadow, which is immaterial. In this case, the marriage of the designer and the user is intimate, as each relies on the other to arrive at the meaning of the designed artifact. For Mendini, a "code" like the stick figure represents a shared resource employed by the designer (in creative work) and the user (through habit) in making sense of designed objects. Every user is therefore a designer, too, engaged in a process of interpretation (semiosis) of codes.[73]

12.6

An "open work": *Stelline* chair (Alessandro Mendini, 1987).

By design, Mendini's whimsical chair, a physical thing, affords the potential for its users to discover a phenomenon outside the material world. The *Stelline* chair and other tangible, designed artifacts that compose Mendini's *oeuvre* are, as defined by this Italian designer, "the linguistic components of an ongoing puzzle that is never completed. The sense lies ... in this expanded, centrifugal movement that has no end. The message of our work lies in this atmospheric dust."[74] The "artificial domestic ecosystem," architectural robotics, is the gathering place for this kind of "atmospheric dust" that promises, through human-machine interactions, pathways to places of social, cultural, and psychological significance.

ARCHITECTURAL ROBOTICS IN THE CONSTELLATION

More than anything else, this book has been concerned with the limits of architecture and computing, striving to envision a cyber-physical environment that supports needs and extends possibilities for human beings. It is far too premature to assess what architectural robotics will become, other than to recognize that there is much more activity in this domain. Something is happening here. Accordingly, to paraphrase the quote from Henri Focillon that appears as the first epigraph at the front of this book, we cannot predict what kind of environment architectural robotics will create, for while it satisfies needs familiar to us, it also precipitates new habits and new opportunities.[75] Whatever the future may bring, implicit in this book is an optimism that comes from an acquaintance with the many promising activities in architecture and computing today, and in the promise they hold, recognizing that architectural robotics is one stream among these currents.

Architectural robotics uses advanced digital technologies to bring us together with others and our physical and digital surroundings, manifested at larger scale, as ecosystems of physical bits, bytes, and biology. Architectural robotics follows from visionary thinking of the past, including Buckminster Fuller's perception of an *Operating Manual for Spaceship Earth* (1968) and Richard Neutra's examination of architecture through the lens of human ecology, *Survival through Design* (1954).[76] In the computing domain, there is the promise of research exemplified by the Augmented Human Assistance (AHA) project just under way at the Madeira Interactive Technologies Institute (Portugal) with participation by an international consortium of academic and industry partners. Paraphrasing from the project's webpage, the design-research team proposes the development and deployment of a novel robotic assistance platform designed to accomplish the following:

Physical (re)training, by way of employing Mixed Reality technology and Serious Games;

Increasing self-awareness, by way of monitoring user-states by means of biosensors, computer vision systems, and exercise performance data, and visualizing this collected information by way of friendly user interfaces, shared with patients, clinicians and/or relatives; and,

Augmented assistance, by way of the integration of the above systems in the form of a mobile, robotic platform with indoor navigation capabilities that will interact, through a virtual coach system, with patients to assist them with their daily task and therapeutic exercises.[77]

Much like the "ecological" design thinking of Fuller and Neutra, the AHA research aims are compelling for striving "to embrace the complexities of the human organism and its community."[78]

While the AHA project is but one, and only recently initiated, there are clear signs that subfields allied to architectural robotics are also aiming to cultivate new vocabularies of design and new, complex realms of understanding toward the realization of novel, human-centered design propositions. Beyond the subfields that this book has made passing references to, tangible computing and physical computing and cyber-physical systems, is the "Internet of Things," which has garnered considerable attention from researchers, industry, and the lay media, and which overlaps, at least partly, with architectural robotics. As defined by the Institute of Electrical and Electronics Engineers (IEEE),

> The Internet of Things (IoT) is a self-configuring and adaptive system consisting of networks of sensors and smart objects whose purpose is to interconnect "all" things, including every day and industrial objects, in such a way as to make them intelligent, programmable and more capable of interacting with humans.[79]

A comparison of architectural robotics and IoT is enabled by a survey of 1,606 technology experts performed and reported by the Pew Research Center.[80] Most obviously, architectural robotics differs from IoT, first, in scale: architectural robotics is localized and intimate; IoT potentially (but not necessarily) reaches everywhere, permeating every environment. (Indeed, 24 billion smart devices are expected to be connected to the Internet by 2016.[81] This makes three smart devices for every human being on the planet.) Second, architectural robotics is not entangled inextricably with the Internet as is central to IoT: architectural robotics can surely benefit from Internet connectivity, but it's envisioned to operate more or less freely from the communication grid as suits its purpose, much as the intelligence of architectural robotics can be allocated and tuned in accord with its purpose. Third, because architectural robotics is, in a physical sense, localized, it has the benefit of being maintained locally (someone nearby knows what technology is present and how to troubleshoot it or at least how to locate the appropriate service to troubleshoot it, much as facilities managers do for building maintenance). Fourth, as architectural robotics is localized, users potentially have more control over their privacy when using the system compared to users of IoT devices, which in many or most cases are exposed to the Internet and subject to the outcomes of raging debates over Internet neutrality.[82] Finally and most significantly, unlike IoT, architectural robotics is concerned with the dimension of (physical) space, providing a capacious home to various devices—an ecosystem tuned to the interactions of people and tools, supporting our life rituals, in a manner that is as old and as meaningful as human civilization.

But the capacity for architectural robotics to help define places of significance comes at a high premium with respect to its cost and its physical scale. Additionally, architectural robotics, IoT, and, in general, robotics share similar challenges: the resource problem of providing battery power to all the devices that compose their systems, as well as the computational challenge of making productive sense of the oceans of data they collect.[83] But arguably for IoT in particular, these are greater challenges for its successful implementation, as the more intimate, localized scale of architectural robotics makes these shared challenges potentially more surmountable. Ultimately, IoT may suffer some from the same issues that smart homes suffer: having a little bit of this technology everywhere, like a skim coat of plaster on the walls of a formidable home. What does it mean, a little bit of this technology everywhere? Do we need and do we want a little bit of intelligence everywhere we look, go, think, relax, play, sleep, learn, heal, create, and dream? Or might we identify where and what kind of this technology belongs in which places within our built environment, and for which purposes, which ultimately is a problem for architects, if they can and have the will to recognize it.[84]

Architectural robotics lends the concept of IoT a physical, dynamic body, the dimension of space, and some needed focus and coherence in the form of localization, fine tuning, and optimization for particular human populations, communities, and terrains that have their own life cycles, needs, and opportunities. Nourbakhsh, for one, envisions a future in which robotic systems are designed "more explicitly for our communities," extending the healthy "dynamic" evident in the Internet and social media "into the physical world we inhabit."[85]

Taking Nourbakhsh a step further, architectural robotics can be conceptualized as a strange and playful "Third Place," especially apt for anchoring community life in an increasingly digital society, facilitating and fostering broader, more creative interaction.[86] In this respect, architectural robotics is much like architecture, rooted locally and connected globally. But unlike architecture, the connectivity of architectural robotics is enhanced by the digital: robotics partly unhinges architecture from its fixed roots. What IoT brings to the foreground of this conclusion is the conception of an ecosystem cohabitated by digital and physical stuff and their hybrids. The Net becomes wider, the digital becomes fleshy, and the material world becomes wired.

John Cage called this kind of place, marvelously, "the garden of technology" where every "inanimate object has a spirit," and where "we" its habitants are more than anything else "undecidable," which is Cage's enigmatic way of saying we are alive.[87] In this rich meadow filled with spirited techno-beings, the limits distinguishing the "organic" and the "technological" have "given way."[88] As architectural theorist Mark Wigley recognizes, drawing on John McHale, "technology itself is an organic system," in which the communication network is a nervous system, architecture is an ecology, and the planet is a machine.[89]

"Ecological design," as David Orr defines it, is "the careful meshing of human purposes with the larger patterns and flows of the natural world and the study of those patterns and flows to inform human actions."[90] Only now, the patterns and flows are plugged in and networked to afford, increasingly, a "complex interweaving of living and nonliving systems" that cultivates a more intimate, responsive, and mutually beneficial corelationship between itself and the *biophilic* people, things, and environmental conditions associated with it.[91] In the not-distant future, we may witness a transition from digital to biological computing, and with it, maybe, the real-time coevolution of architectural robotics and its inhabitants that marks a return "to a more organic, integral way of being, to restore, in a way, the holistic world ... that earlier technologies dislocated us from."[92] An expanding network of architectural robotic ecosystems promises to dignify, deepen, and challenge our humanity and will demand that people and robots of all kinds live well together.

"All Watched Over by Machines of Loving Grace"
—Richard Brautigan

I like to think (and
the sooner the better!)
of a cybernetic meadow
where mammals and computers
live together in mutually
programming harmony
like pure water
touching clear sky.

I like to think
(right now, please!)
of a cybernetic forest
filled with pines and electronics
where deer stroll peacefully
past computers
as if they were flowers
with spinning blossoms.

I like to think
(it has to be!)
of a cybernetic ecology
where we are free of our labors
and joined back to nature,
returned to our mammal
brothers and sisters,
and all watched over
by machines of loving grace.

From *All Watched Over by Machines of Loving Grace*
by Richard Brautigan. Copyright ©1967 by Richard Brautigan.
Reprinted with the permission of the Estate of Richard Brautigan,
all rights reserved.

NOTES

CHAPTER ONE

1. Stuart K. Card, Thomas P. Moran, and Allen Newell, *The Psychology of Human-Computer Interaction* (Hillsdale, N.J.: Lawrence Erlbaum Associates, 1983). The term *human-computer interaction* reportedly first appeared in James H. Carlisle, "Evaluating the Impact of Office Automation on Top Management Communication," in *American Federation of Information Processing Societies Conference Proceedings. Volume 45. National Computer Conference, June 7–10, 1976, New York* (Montvale, N.J.: AFIPS Press, 1976), 611–616.

2. M. Weiser, "The Computer for the 21st-Century," *Scientific American* 265, no. 3 (1991): 94.

3. Ibid.

4. S. Shafer, "Ten Dimensions of Ubiquitous Computing," in *Managing Interactions in Smart Environments*, ed. P. Nixon, G. Lacey, and S. Dobson (London: Springer-Verlag, 2000), 5–16.

5. Sherry Turkle, *Alone Together: Why We Expect More from Technology and Less from Each Other* (New York: Basic Books, 2011).

6. T. Someya, "Building Bionic Skin," *IEEE Spectrum* 50, no. 9 (2013): 50–56.

7. Robin Marantz Henig, "The Real Transformers," *New York Times Magazine* 156, no. 54020 (2007): 28–55.

8. More precisely, the *singularity* has been defined as an "intelligence explosion" of one or another form: either machine intelligence comes to meet or exceed human intelligence or technologically amplified human intelligence comes to exceed human intelligence to such an extent that it defines a post-human race. An overview of the singularity, its definitions, and the energetic debate surrounding it is found in Ammon H. Eden, Eric Steinhart, David Pearce, and James H. Moor, "Singularity Hypotheses: An Overview," in *Singularity Hypotheses: A Scientific and Philosophical Assessment*, ed. James H. Moor, Ammon H. Eden, Johnny H. Soarker, and Eric Steinhart (New York: Springer, 2012), 1–12. We will return to the singularity later in this book.

9. I am drawing on the definition of cyber-physical systems (of which architectural robotics can arguably be counted as a stream) as offered by Imre Horváth and Bart H. M. Gerritsen, "Cyber-Physical Systems: Concepts, Technologies and Implementation Principles," in *TMCE 2012, the International Symposium on Tools and Methods of Competitive Engineering*, ed. Z. Rusák, I. Horváth, A. Albers, and M. Behrendt (Karlsruhe, Germany, 2012), 19–36. As defined by Horváth and Gerritsen, "in the broadest sense, cyber-physical systems (CPSs) blend the knowledge and technologies of the third wave of information processing, communication and computing with the knowledge and technologies of physical artifacts and engineered systems" (p. 19). In their "synergic model" of CPSs, Horváth and Gerritsen recognize physical technologies, cyber technologies, and their hybridization as cyber-physical systems as a single "formation" that collectively shapes the "human domain" and the "socio-techno-economic environment domain" (p. 21). We will return to CPSs later in this book.

CHAPTER TWO

1. Harry Francis Mallgrave, *The Architect's Brain: Neuroscience, Creativity, and Architecture* (Chichester, UK: Wiley-Blackwell, 2010), 176. The word *metaphor* literally means "a carrying over" in both the physical and the figurative sense. In metaphor, words are "carried over" limits or "transferred" to new contexts. Nigel Lewis, while indicating that there are some 125 definitions of metaphor, characterized metaphor as an "abstraction or extension from the supposedly commonplace" whereby "words break out of the careful categories of etymology into a broad field of perception" (Nigel Lewis, *The Book of Babel: Words and the Way We See Things* [Iowa City: University of Iowa Press, 1994], 4).

2. Gerald M. Edelman, *Second Nature: Brain Science and Human Knowledge* (New Haven, Conn.: Yale University Press, 2006), 58–59. Following the same line of research, it has been hypothesized by Lakoff and Johnson that many metaphors (e.g., a "warm" or "cold" personality) are hard-wired into neural maps at the earliest stage of human development (see George Lakoff and Mark Johnson, *Philosophy in the Flesh: The Embodied Mind and Its Challenge to Western Thought* [New York: Basic Books, 1999]).

3. José Ortega y Gasset, "La Deshumanizacíon del arte e Ideas sobre la Novella" [1925], in *The Dehumanization of Art and Other Essays on Art, Culture and Literature*, trans. H. Weyl (Princeton, N.J.: Princeton University Press, 1972), 33.

4. Don Ihde contends that theaters, dating back to "Plato's cave-theatre of 2400 years ago," are not only "artifices" but also "'technological set-ups'" and what he calls "epistemology engines" for gaining "knowledge and human experience." In a tradition Ihde identifies across theater, film, and computing that includes Plato, Descartes, Locke, and *The Matrix* film trilogy, "some technology or technology complex provides the model for knowledge and human experience"; that is, "in its latest incarnation ... a differently shaped theatre" (Don Ihde, *Embodied Technics* [Copenhagen: Automatic Press, 2010]).

5. Aldo Rossi, *A Scientific Autobiography* (Cambridge, Mass.: MIT Press, 1981), 33.

6. Ibid.

7. Ibid.

8. Ibid.

9. Gray Read, "Theater of Public Space: Architectural Experimentation in the Théâtre De L'espace (Theater of Space), Paris 1937," *Journal of Architectural Education* 58, no. 4 (2005): 2. Autant and Lara's conceptualization of the theater had much in common with the theater experiments of Italian Futurism and Russian Constructivism that mostly preceded their activities; however, Autant and Lara most explicitly used theater as a generator for architectural design.

10. Norbert Weiner, "The Machine Age" [1949], excerpt in John Markoff, "In 1949, He Imagined an Age of Robots," *New York Times* (May 21, 2013): D8. Markoff describes the thwarted effort of the *New York Times* to publish Weiner's visionary article of 1949 that summarized his thinking, following his landmark *Cybernetics* published one year earlier and the *Times*' subsequent request of Weiner to summarize his views in an essay about "what the ultimate machine age is likely to be." Weiner's essay only made it to print in 2013, and only in excerpt, within Markoff's article as cited here.

11. Gordon Pask, "The Architectural Relevance of Cybernetics," *Architectural Design* (September 1969): 494–496.

12. Ibid., 495.

13. Ibid.

14. Ibid., 496.

15. Ibid.

16. Ibid.

17. Ibid., 495.

18. Christopher Alexander, Sara Ishikawa, and Murray Silverstein, *A Pattern Language: Towns, Buildings, Construction* (New York: Oxford University Press, 1977), xv and x.

19. Ibid., 858–859.

20. Ibid.

21. Ibid., 1166.

22. In *Elements of Reusable Object-Oriented Software*, Gamma, Helm, Johnson, and Vlissides construct their argument for pattern-based software design as inspired by Alexander's concept of recurring problems in the environment that suggest a pattern-based solution that can be reinterpreted in countless ways. See Erich Gamma, Richard Helm, Ralph Johnson, and John Vlissides, *Design Patterns: Elements of Reusable Object-Oriented Software* (Reading, Mass.: Addison-Wesley Longman, 1995). From this broader consideration of the utility of Alexander's pattern recognition to computer science come more specialized applications, as in computer game design (Bernd Kreimeier, "The Case for Game Design Patterns," available at http://echo.iat.sfu.ca/library/kreimeier_02_game_patterns.pdf).

23. A consideration of the utility of Alexander's *A Pattern Language* for human-robot interaction (HRI) is found in Peter H. Kahn et al., "Design Patterns for Sociality in Human-Robot Interaction," in *Proceedings of the 3rd ACM/IEEE International Conference on Human Robot Interaction* (Amsterdam, Netherlands: ACM, 2008), 97–104.

24. Alexander, Ishikawa, and Silverstein, *A Pattern Language*, xliii.

25. Designed by Gary Chang of EDGE Design Institute, the Domestic Transformer has been alternatively referred to as the "Hong Kong Space Saver" (see https://www.youtube.com/watch?v=f-iFJ3nclDoand www.edgedesign.com).

26. Alexander, Ishikawa, and Silverstein, *A Pattern Language*, xliii–xliv.

27. Ibid., xliii.

28. Recognize that Alexander never insisted that the patterns of *A Pattern Language* represented absolute relationships between people, things, and their immediate physical environment, to be followed slavishly by designers; rather, he characterized the patterns as "very much alive and evolving," affected by "new experience and observation" (Alexander, Ishikawa, and Silverstein, *A Pattern Language*, xv). The same can be said, in concept, for architectural robotics.

29. On actor-network theory and the "Internet of Things," see, for instance, J. S. Dolwick, "'The Social' and Beyond: Introducing Actor-Network Theory," *Journal of Maritime Archaeology* 4, no. 1 (2009): 21–49; Adrian McEwen and Hakim Cassimally, *Designing the Internet of Things* (West Sussex, UK: Wiley, 2014).

30. The interface of architecture and computing is formally strong at MIT: its highly visible Media Lab, founded by Negroponte and Jerome Wiesner, emerged from the Architecture Machine Group and remains within the School of Architecture. William Mitchell was later dean of the School of Architecture and also served as head of academic programs (Media Arts and Sciences) of the Media Lab.

31. Nicholas Negroponte, *Soft Architecture Machines* (Cambridge, Mass.: MIT Press, 1975), 135.

32. Ibid., 128 and 134.

33. Ibid., 135 and 128.

34. Ibid., 126.

35. Ibid., 129.

36. Ibid.

37. Ibid., 128.

38. Ibid., 127.

39. Ibid.

40. Ibid., 132.

41. Ibid., 145.

42. William J. Mitchell, *E-Topia: Urban Life, Jim—but Not as We Know It* (Cambridge, Mass.: MIT Press, 1999), 59.

43. H. W. J. Rittel and M. M. Webber, "Dilemmas in a General Theory of Planning," *Policy Sciences* 4, no. 2 (1973): 155–169. The term *wicked problems* was first used by C. West Churchman, "Guest Editorial: Wicked Problems," *Management Science* 14, no. 4 (1967): 141–142.

44. J. Zimmerman, J. Forlizzi, and J. Evenson, "Research through Design as a Method for Interaction Design Research in HCI." *Proceedings of CHI '07, the ACM Conference on Human Factors in Computing Systems* (2007): 493–502.

45. Ibid., 2 and 6.

46. Ibid., 6. Different names have been given to the design research process described here (partly dependent on which design community is engaged in the research), including human-centered design, participatory design, critical design, reflective design, and evidence-based design. These terms are not all interchangeable but are closely related. For one, Zimmerman and colleagues refer to the design research process as "research through design," a terminology drawn from C. Frayling (C. Frayling, "Research in Art and Design," *Royal College of Art Research Papers* 1, no. 1 [1993]: 1–5) that defines an emerging subfield within HCI, as evidenced by ACM conference sessions dedicated to it.

47. Ibid., 3.
48. P. J. Stappers, "Designing as a Part of Research," in *Design and the Growth of Knowledge: Best Practices and Ingredients for Successful Design Research*, ed. R. van der Lugt and P. J. Stappers (Delft, Netherlands: ID Studiolab Press, 2006), 12–17.
49. Ibid., 14.
50. Negroponte, *Soft Architecture Machines*.
51. Gaston Bachelard, *The Poetics of Space*, trans. M. Jolas (New York: Beacon Press, 1969), 64. In what is arguably Bachelard's most well-known work, he praises the "poet who follows the working draft of … metaphors to build his house."
52. Weiner, "The Machine Age."
53. G. Raz (producer). *TED Radio Hour: Framing the Story* (audio podcast, June 7, 2013). Available at www.apple.com/itunes/.
54. C. Rose (producer). *Charlie Rose: Margaret Atwood* (video podcast, October 7, 2013). Available at www.charlierose.com.
55. CHI is short for the Conference on Human Factors in Computing Systems, which is hosted by the Association for Computer Machinery, the world's largest educational and scientific computing society (see www.sigchi.org/conferences and www.acm.org). The abstract for the workshop cited here was published as Conor Linehan et al., "Alternate Endings: Using Fiction to Explore Design Futures," in *CHI '14 Extended Abstracts on Human Factors in Computing Systems* (Toronto: ACM, 2014): 45–48.
56. Ibid., 45.
57. Nathan et al. are referenced in Ibid., 47 and 48, note 12.
58. Ibid., 47.

CHAPTER THREE
1. Henri Focillon, *The Life of Forms in Art* (New York: Zone Books, 1989), 34–35; see also 97–98.
2. Michel de Certeau, *The Practice of Everyday Life* (Berkeley, Calif.: University of California Press, 1984), 201.
3. Focillon, *The Life of Forms in Art*, 60.
4. Ibid., 123–124.
5. A consideration of the medieval process of grafting is found in Robert Odell Bork, William W. Clark, and Abby McGehee, *New Approaches to Medieval Architecture*, Avista Studies in the History of Medieval Technology, Science and Art (Burlington, Vt.: Ashgate, 2011), 201.
6. For a consideration of these kinds of intelligent car technologies, see Aaron M. Kessler, "Technology Takes the Wheel," *New York Times* (October 6, 2014): B1. For intelligent technology specifically for drivers falling asleep at the wheel, see "Germans Develop Device to Stop Drivers Falling Asleep at the Wheel," *Telegraph* (November 10, 2010). Available at www.telegraph.co.uk/technology/news/8122957/Germans-developdevice-to-stop-drivers-falling-asleep-at-the-wheel.html.

7. There are wonderful exceptions to architecture's petrified tendencies, such as the concept of "event scape" and "hybrid programming" offered by Swiss architect Bernard Tschumi (see, for instance, Bernard Tschumi, *Event-Cities: Praxis* (Cambridge, Mass.: MIT Press, 1994); *Architecture and Disjunction* (Cambridge, Mass.: MIT Press, 1994).
8. Leon Battista Alberti, *On the Art of Building in Ten Books*, trans. Joseph Rykwert, Neil Leach, and Robert Tavernor (Cambridge, Mass.: MIT Press, 1988), book VI, c.2, 156.
9. Ibid.
10. The term *affordances* and its variants are used frequently in this book. Psychologist James J. Gibson introduced the term in the late 1970s to refer to *action possibilities* within an artifact or environment (e.g., a door swings due to its hinges), independent of whether an individual in that environment can perceive these possibilities or not. In the late 1980s, Donald Norman appropriated the term for the HCI and interaction design communities to refer to those *action possibilities* within an artifact or environment that are both *readily perceivable* by an actor and likely to define the interaction between actor and artifact. Norman's definition suggests an "ecological approach" in which affordances are *relational*—an interaction between the artifact and the individual interacting with it—rather than only subjective or only intrinsic to the artifact. This book uses the ecological connotation of *affordance* offered by Norman. See James J. Gibson, *The Ecological Approach to Visual Perception* (Boston: Houghton Mifflin, 1979); and from Donald A. Norman, *The Psychology of Everyday Things* (New York: Basic Books, 1988) and *The Design of Everyday Things* (New York: Basic Books, 2013).
11. Carlo Mollino, "Utopia e Ambientazione," *Domus* (August 1949): 16 (translation by author Keith Evan Green). For an overview of flexible furniture, see Sigfried Giedion, *Mechanization Takes Command, a Contribution to Anonymous History* (New York: Oxford University Press, 1948); and Phyllis Bennett Oates, *The Story of Western Furniture* (London: Herbert Press, 1981).
12. Umberto Eco, *The Open Work*, trans. A. Cancogni (Cambridge, Mass.: Harvard University Press, 1989), 4.
13. Ibid., 163.
14. Gilles Deleuze and Félix Guattari, *Nomadology: The War Machine* (New York: Semiotext(e), 1986), 17.
15. Steven Levy, *Artificial Life: A Report from the Frontier Where Computers Meet Biology* (New York: Vintage Books, 1993), 18.
16. Homer, *The Odyssey*, trans. Ian C. Johnston (Arlington, Va.: Richer Resources Publications, 2007).
17. For more on the many and variegated forms named Proteus, see Senthy V. Sellaturay, Raj Nair, Ian K. Dickinson, and Seshadri Sriprasad, "Proteus: Mythology to Modern Times," *Indian Journal of Urology* 28, no. 4 (2012). Available at www.ncbi.nlm.nih.gov/pmcarticles/PMC3579116.
18. *Oxford English Dictionary* (available at www.oed.com/view/Entry/153124?redirectedFrom=protean#eid) and *Oxford Dictionaries* (available at www.oxforddictionaries.com/us/definition/american_english/protean?q=protean).

19. Robert Jay Lifton, *The Protean Self: Human Resilience in an Age of Fragmentation* (New York: Basic Books, 1994), 1.

20. Ibid., 14–15.

21. Susan C. Antón, Richard Potts, and Leslie C. Aiello, "Evolution of Early Homo: An Integrated Biological Perspective," *Science* 345, no. 6192 (2014): 1236828, 8.

22. Ibid.

23. Ibid., 1236828, 9 and 1236828, 7.

24. Ibid., 1236828, 7.

25. Lifton, *The Protean Self*, 230.

CHAPTER FOUR

1. Richard A. Shweder, "Keep Your Mind Open (and Watch It Closely, Because It's Apt to Change)," *New York Times Books in Review* (February 20, 1994): 16.

2. Nikil Saval, *Cubed: A Secret History of the Workplace* (New York: Doubleday, 2014). More profound but less focused on the office workspace and the nature of office work is *Mechanization Takes Command* (1948), Sigfried Giedion's examination of mechanization and its effects on everyday life, referenced elsewhere in this book.

3. These are Jill Lepore's words from her review of Saval's book. See Jill Lepore, "Away from My Desk," *New Yorker* (May 12, 2014): 76.

4. T. W. Malone, "How Do People Organize Their Desks? Implications for the Design of Office Information Systems," *ACM Transactions on Office Information Systems* 1, no. 1 (1983): 99–112.

5. See O. Bondarenko and R. Janssen, "Documents at Hand: Learning from Paper to Improve Digital Technologies," in *Proceedings of CHI '05, the ACM Conference on Human Factors in Computing Systems* (New York: ACM, 2005), 121–130; and Abigail J. Sellen and Richard H. R. Harper, *The Myth of the Paperless Office* (Cambridge, Mass.: MIT Press, 2002).

6. The elaboration of AWE over the next pages draws from the publications of the author and his research team. Key among these publications is the following one, which provides the overall arc of the research, from inception through evaluation: Henrique Houayek et al., "AWE: An Animated Work Environment for Working with Physical and Digital Tools and Artifacts," *Personal and Ubiquitous Computing* 18 (2014): 1227–1241.

7. Some examples from this literature include the following:
   - On the use of multiple displays, D. Wigdor et al., "Effects of Display Position and Control Space Orientation on User Preference and Performance," in *Proceedings of CHI '06, the ACM Conference on Human Factors in Computing Systems* (New York: ACM, 2005), 309–318; and R. Ziola, "My MDE: Configuring Virtual Workspace in Multidisplay Environments," in *Proceedings of CHI '05, the ACM Conference on Human Factors in Computing Systems, Extended Abstracts (Work in Progress)* (2006), 1481–1486.

- On managing mixed media, P. Luff et al., "Handling Documents and Discriminating Objects in Hybrid Spaces," in *Proceedings of CHI '06, the ACM Conference on Human Factors in Computing Systems* (2006), 561–570.
- On viewing health-care information, U. Varshney, "Pervasive Healthcare,"*Computer* 36, no. 12 (2003): 139–140.
- On computer-supported collaborative work (CSCW), Ronald M. Baecker, *Readings in Groupware and Computer Supported Cooperative Work* (San Mateo, Calif.: Morgan Kaufmann, 1993).

8. Some examples from this literature include the following:
- On the design of work environments, see Paola Antonelli, *Workspheres: Design and Contemporary Work Styles* (New York: Museum of Modern Art and Harry N. Abrams, 2001).
- On the Interactive Workspaces Project, see B. Johanson, A. Fox, and T. Winograd, "The Interactive Workspaces Project: Experiences with Ubiquitous Computing Rooms," *IEEE Pervasive Computing* 1, no. 2 (2002): 67–74.
- On Roomware, see N. A. Streitz et al., "Roomware: Toward the Next Generation of Human-Computer Interaction Based on an Integrated Design of Real and Virtual Worlds," in *Human–Computer Interaction in the New Millennium*, ed. J. Carroll (Boston: Addison-Wesley, 2001), 553–578.

9. While a few years old now, *Workspheres*, the Museum of Modern Art exhibition and catalog (identified in the previous note), is exemplary for considering the high design of workspaces.

10. Giovanna Borradori, "Recoding Metaphysics: Strategies of Italian Contemporary Thought," in *Recoding Metaphysics: The New Italian Philosophy*, ed. Giovanna Borradori (Evanston, Ill.: Northwestern University Press, 1988), 5. See also Giovanna Borradori, "'Weak Thought' and Post Modernism: The Italian Departure from Deconstruction," *Social Text* 18 (Winter 1987/88): 39–49.

11. What is described here is the law of *prägnanz* (German for "pithiness"), the fundamental principle of gestalt perception. See Wolfgang Köhler, *Gestalt Psychology: An Introduction to New Concepts in Modern Psychology* (New York: Liveright, 1992).

12. See, for instance, Brigitte Jordan, "Pattern Recognition in Human Evolution and Why It Matters for Ethnography, Anthropology, and Society," in *Advancing Ethnography in Corporate Environments: Challenges and Emerging Opportunities*, ed. Brigitte Jordan (Walnut Creek, Calif.: Left Coast Press, 2013).

13. For an overview of pattern recognition theory in cognitive psychology, see, for instance, Youguo Pi, Wenzhi Liao, Mingyou Liu, and Jianping Lu, "Theory of Cognitive Pattern Recognition," in *Pattern Recognition Techniques, Technology and Applications*, ed. Peng-Yeng Yin (Vienna: InTech, 2008), 433–462. For insight into how pattern recognition is being considered in neuroscience and specifically in the area of neural networks, see, for instance, Bart Kosko, "Bidirectional Associative Memories," *IEEE Transactions on Systems, Man and Cybernetics* 18, no. 1 (1988): 49–60.

14. L. Rutkowski, "On Bayes Risk Consistent Pattern Recognition Procedures in a Quasi-Stationary Environment," *IEEE Transactions on Pattern Analysis and Machine Intelligence* 4, no. 1 (1982): 84–87.

15. Erik Grönvall et al., "Causing Commotion with a Shape-Changing Bench: Experiencing Shape-Changing Interfaces in Use," in *Proceedings of CHI '14, the ACM Conference on Human Factors in Computing Systems* (New York: ACM, 2014), 2559–2568. For related work on shape-shifting furniture, see by the same group, Majken K. Rasmussen et al., "Shape-Changing Interfaces: A Review of the Design Space and Open Research Questions," in *Proceedings of the CHI' 12, the ACM Conference on Human Factors in Computing Systems* (Austin, Tex.: ACM, 2012), 735–744; and by Matthijs Kwak et al., "The Design Space of Shape-Changing Interfaces: A Repertory Grid Study," in *Proceedings of the CHI '14, the ACM Conference on Human Factors in Computing Systems* (ACM, 2014), 181–190.

16. For more on the technical design of *coMotion*, see Sofie Kinch et al., "Encounters on a Shape-Changing Bench: Exploring Atmospheres and Social Behaviour in Situ," in *Proceedings from TEI '14, the International Conference on Tangible, Embedded and Embodied Interaction* (New York: ACM, 2013), 233–240.

17. Bruce Mau, "An Incomplete Manifesto for Growth," in *Life Style*, ed. Bruce Mau, Bart Testa, and Kyo Maclear (London: Phaidon, 2000).

18. WoZ techniques are used widely in HCI and related research. For a consideration of their use in computational artifacts more akin to AWE, see, for instance, S. Dow et al., "Wizard of Oz Interfaces for Mixed Reality Applications," in *Proceedings of CHI '05, the ACM Conference on Human Factors in Computing Systems, Extended Abstracts* (New York: ACM, 2005), 1339–1342.

19. The development of these guidelines is elaborated in K. E. Green et al., "Configuring an 'Animated Work Environment': A User-Centered Design Approach," paper presented at IE 2008, the 4th International Conference on Intelligent Environments, Seattle, Washington, July 21–22, 2008.

20. Jonah Lehrer, "Groupthink: The Brainstorming Myth," *New Yorker* (January 30, 2012), 22–27.

21. See Bondarenko and Janssen, "Documents at Hand"; Sellen and Harper, *The Myth of the Paperless Office*; and also A. Kidd, "The Marks Are on the Knowledge Worker," in *Proceedings of CHI '94, the ACM Conference on Human Factors in Computing Systems* (New York: ACM, 1994), 186–191.

22. See "Robots Make Better Employees Than People, America's Business Leaders Say." Available at http://www.hngn.com/articles/41898/20140909/americas-business-leaders-hire-robots-over-people-survey-finds.htm.

23. The term *robot* was first introduced by the Czech playwright Karel Capek in his 1920 play *Rossum's Universal Robots,* the word *robota* being the Czech word for "forced labor" or "slave" (see Mark W. Spong, Seth Hutchinson, and M. Vidyasagar, *Robot Modeling and Control* [Hoboken, N.J.: John Wiley & Sons, 2006], 1).

24. Even in such structured environments, robots have caused dozens of injuries and deaths in the United States. See John Markhoff and Claire Cain Miller, "Danger: Robots Working," *New York Times* (June 17, 2014): D1 and D5.

25. Hans Moravec, quoted (from a text of 1988) in John Markoff, "Brainy, Yes, but Far from Handy," *New York Times* (September 1, 2014): D1.

26. The topic of when or if superintelligent humanoid robots will arrive is a topic of fierce debate among members of faculty and industry inside and allied to robotics and will be considered again in the epilogue of this book. For the moment, I reference one such debate that I attended in a meeting of the Robotics Institute of the Technical University of Delft (TU Delft) on October 31, 2014. Approximately one hundred attended the meeting, representing robotics researchers from across the university, and only some half of them voted that such a robot would be realized in our lifetimes. For more on this meeting and the Robotics Institute of TU Delft, see www.robotics.tudelft.nl/?q=news/recap-event-311014-next-phase.

27. Chapter 6 in *Making Things See*, from the popular *Make:Books* series, offers an overview of kinematics and instructions for the robotics hobbyist to learn kinematics firsthand by programming a simple robotic arm in and with Arduino (see Greg Borenstein, *Making Things See* [Sebastopol, Calif.: Maker Media, 2012]). For a more sophisticated but still introductory treatment of kinematics and the field of robotics overall, see chapter 14, "Robotics," in Frank Kreith and D. Yogi Goswami, *The CRC Handbook of Mechanical Engineering*, 2nd ed., 2 vols., Mechanical Engineering Handbook Series (Boca Raton, Fla.: CRC Press, 2005). A more thorough introduction yet to kinematics is found in C. S. George Lee, "Robot Arm Kinematics, Dynamics, and Control," *Computer* 15, no. 12 (1982): 62–80.

28. This is an adequate definition for a robot; however, as might be expected, robots have been defined in many other ways, which results in some confusion about what a robot is and does. "Virtually anything that operates with some degree of autonomy, usually under computer control," argue Spong, Hutchinson, and Vidyasagar, "has at some point been called a robot" (Spong, Hutchinson, and Vidyasagar, *Robot Modeling and Control*, 1). Despite this confusion, the same authors, emphasizing the functional aspect of robotics, define a robot as "a reprogrammable, multifunctional manipulator designed to move materials, parts, tools, or specialized devices through variable programmed motions for the performance of a variety of tasks" (p. 2). While striving to redefine *robot* for an emerging, expanding field of robotics, James Trevelyan recognizes that any definition of a robot only limits the number and kind of robots that might be developed and used (James Trevelyan, "Redefining Robotics for the New Millennium," *International Journal of Robotics Research* 18 [1999]: 1221). Compelled to offer a definition in any case, Trevelyan offers, rather simply, that the field of robotics is the "science of extending human motor capabilities with machines" (p. 1222). All three of the citations made here serve as fine resources for learning more about robotics, as does the more recent book by Maja J. Mataric, *The Robotics Primer* (Cambridge, Mass.: MIT Press, 2007).

29. Much of this paragraph and the ones that follow in this section draw on publications by my lab: on the technical design of AWE, M. Kowka et al., "The AWE Wall: A Smart Reconfigurable Robotic Surface," paper presented at IE 2008, the 4th International Conference on Intelligent Environments, Seattle, Washington, July 21–22, 2008; and on the design and testing of AWE, Houayek et al., "AWE: An Animated Work Environment for Working with Physical and Digital Tools and Artifacts."

30. Here, I am evoking something of a cyber-physical manifestation of the concept of "third places," that social environment that lies outside the home and the workplace. See Ramon Oldenburg and Dennis Brissett, "The Third Place," *Qualitative Sociology* 5 no. 4 (1982): 265–284.

CHAPTER SIX

1. Edgar Allan Poe, *The Fall of the House of Usher, and Other Tales* (New York: New American Library, 1960).

2. Ibid.

3. Anthony Vidler, *The Architectural Uncanny: Essays in the Modern Unhomely* (Cambridge, Mass.: MIT Press, 1992), 15–44.

4. Giorgio de Chirico, *Hebdomeros*, trans. John Ashbery (Cambridge, Mass.: Exact Change, 1992).

5. de Chirico, "Statues, Furniture and Generals," in *Hebdomeros*, 244–245.

6. Harry Francis Mallgrave, *The Architect's Brain: Neuroscience, Creativity, and Architecture* (Chichester, UK: Wiley-Blackwell, 2010), 58.

7. Immanuel Kant, *Critique of Pure Reason*, ed. Jim Manis, trans. J. M. D. Meiklejohn, Dover Philosophical Classics (Hazleton, Pa.: The Electronics Classic Series, 2013), 14.

8. Mallgrave, *The Architect's Brain*, 159. For Mallgrave, "neuroscience is reminding us of the enormous complexities of what were once viewed as simply sensory reflexes to stimuli" (p. 159).

9. "Omnisensorial" suggests the effects of light, sound, texture, color, scent, dampness, and resiliency of/on an architectural work that affect one's experience of it. For more on the omnisensorial qualities of architecture, see Richard Joseph Neutra, *Survival through Design* (New York: Oxford University Press, 1954).

10. Aristotle, *Generation of Animals*, trans. A. L. Peck, Loeb Classical Library (Cambridge, Mass.: Harvard University Press, 1943), book II, ch. 1, 145.

11. Ibid., 149.

12. It should be noted that Goethe, and not Aristotle, was likely the first to use the term *morphology* to describe the science of forms. In 1800, immediately following Goethe, the anatomist Karl-Friedrich Burdach used the word in what is considered its first appearance in print.

13. Aristotle, *Generation of Animals*, 151.

14. Lisa Nocks, *The Robot: The Life Story of a Technology* (Westport, Conn.: Greenwood Press, 2007).

15. Kenneth Gross, *The Dream of the Moving Statue* (Ithaca, N.Y.: Cornell University Press, 1992), xi.

16. Joseph Rykwert, "Organic and Mechanical," *RES* 22 (1992): 13.

17. John von Neumann, "The General and Logical Theory of Automata," in *John Von Neumann: Collected Works*, ed. A. H. Taub (Elmsford, N.Y.: Pergamon, 1963), vol. 5, 288–328.

18. Ibid.

19. For an overview of the a-life (artificial-life community), see Steven Levy, *Artificial Life: A Report from the Frontier Where Computers Meet Biology* (New York: Vintage Books, 1993), 6. For a lengthy introduction to the artistic works of Theo Jansen, see the online version of Lawrence Weschler, "Theo Jansen's Lumbering Life-Forms Arrive in America," *New York Times Magazine* (November 26, 2014). Available at http://www.nytimes.com/2014/11/30/magazine/theo-jansens-lumbering-life-forms-arrive-in-america.html?_r=1/.

20. John McHale, *The Future of the Future* (New York: G. Braziller, 1969), 99.

21. Sigfried Giedion, *Mechanization Takes Command, a Contribution to Anonymous History* (New York: Oxford University Press, 1948), 7.

22. Ibid., 271.

23. Ibid., 273.

24. Ibid., 14.

25. Ibid., 15.

26. Ibid., 264.

27. Ibid., 312.

28. Ibid., 316–317.

29. Ibid., 316.

30. Ibid.

31. Ibid., 328.

32. Ibid., 317. On the latter, Giedion wrote, "One does not have to consult the late eighteenth-century engravings to sense the erotic atmosphere of these apartments" (p. 317).

33. Ibid., 318–319.

34. Ibid., 484.

35. Ibid., 25.

36. Ibid., 312.

37. Erico Guizzo, "So, Where Are My Robot Servants?: Tomorrow's Robots Will Become True Helpers and Companions In People's Homes—And Here's What It Will Take to Develop Them," *IEEE Spectrum*, June (2014): 74.

CHAPTER SEVEN

1. See T. Fong, I. Nourbakhsh, and K. Dautenhahn, "A Survey of Socially Interactive Robots," *Robotics and Autonomous Systems* 42 (2003): 143–166.

2. K. Park et al., "Robotic Smart House to Assist People with Movement Disabilities," *Autonomous Robot* 22 (2007): 183–198.

3. Terry Winograd, *Bringing Design to Software* (Reading, Mass.: Addison-Wesley, 1996), 165.
4. This aspect of the research is elaborated in A. L. Threatt et al., "A Vision of the Patient Room as an Architectural-Robotic Ecosystem," in *IROS 2012, the IEEE/RJS International Conference on Intelligent Robots and Systems* (New York: IEEE, 2012), 3223–3224.
5. Donald A. Norman, "The Next UI Breakthrough, Part 2: Physicality," *Interactions*, July and August (2007): 46–47.
6. Nicholas Negroponte, *Soft Architecture Machines* (Cambridge, Mass.: MIT Press, 1975), 127–128.
7. Malcolm McCullough, *Digital Ground: Architecture, Pervasive Computing, and Environmental Knowing* (Cambridge, Mass.: MIT Press, 2004), xx.
8. P. M. Yanik et al., "A Gesture Learning Interface for Simulated Robot Path Shaping with a Human Teacher," *IEEE Transactions on Human Machine Systems* 44, no. 1 (2013): 41–54; and A. L. Threatt et al., "Design and Evaluation of a Nonverbal Communication Platform for Human-Robot Interaction," in *Proceedings of HCI International 2013*, ed. N. Streitz and C. Stephanidis, Springer Lecture Notes in Computer Science (Heidelberg: Springer, 2013), 505–513.
9. See, for instance, K. Dautenhahn, "Socially Intelligent Robots: Dimensions of Human-Robot Interaction," *Philosophical Transactions of the Royal Society of London, Series B: Biological Sciences* 362, no. 1480 (2007): 679–704; B. Mutlu et al., "Social Robotics," paper presented at the Proceedings of ICSR 2011, the Third International Conference on Social Robotics, Amsterdam, November 24–25, 2011; D. S. Syrdal et al., "The Negative Attitudes Towards Robots Scale and Reactions to Robot Behaviour in a Live Human Robot InterAction Study," paper presented at the Proceedings on New Frontiers in Human- Robot Interaction, AISB 2009 Convention, Edinburgh, Scotland, April 6–9, 2009; and T. Komatsu, "Audio Subtle Expressions Affecting User's Perceptions," in *Proceedings of IUI 2006, the International Conference on Intelligent User Interface* (New York: ACM, 2006), 306–308.
10. R. Read and T. Belpaeme, "Interpreting Non-Linguistic Utterances by Robots: Studying the Influence of Physical Appearance," in *Proceedings of AFFINE '10, the 3rd International Workshop on Affective Interaction in Natural Environments* (New York: ACM, 2010), 65–70.
11. "Pain in Non-Verbal Elderly Largely Undetected by Family Caregivers," available at http://m2.facebook.com/americanpainsociety.
12. D. Quenqua, "Pushing Science's Limits in Sign Language Lexicon," *New York Times* (December 4, 2012): D1.
13. J. Cassell, "A Framework for Gesture Generation and Interpretation," in *Computer Vision for Human–Machine Interaction*, ed. R. Cipolla and A. Pentland (Cambridge: Cambridge University Press, 1998), 191–215.
14. Threatt et al., "Design and Evaluation of a Nonverbal Communication Platform for Human-Robot Interaction," 505–513. This segment of the chapter draws generously from this published paper authored by our team.
15. Negroponte, *Soft Architecture Machines*, 127.

16. ART's gesture-recognition system is elaborated in our technical paper cited earlier in this chapter (from which this segment draws generously): Yanik et al., "A Gesture Learning Interface for Simulated Robot Path Shaping with a Human Teacher," 41–54.

17. Indeed, hand-and-arm gesticulation accounts for some 90 percent of gestured communication. See S. Mitra and T. Acharya, "Gesture Recognition: A Survey," *IEEE Transactions on Systems, Man and Cybernetics, Part C, Applications and Reviews* 37, no. 3 (2007): 311–324.

18. For a recent and thorough consideration of vision-based sensor platforms and competing sensor types for the purpose of gesture recognition, see S. Berman and H. Stern, "Sensors for Gesture Recognition Systems," *IEEE Transactions on Systems, Man, and Cybernetics, Part C: Applications and Reviews* 42, no. 3 (2012): 277–290.

19. Yanik et al., "A Gesture Learning Interface for Simulated Robot Path Shaping with a Human Teacher," 43.

20. Ray Kurzweil, *The Singularity Is Near: When Humans Transcend Biology* (New York: Viking, 2005). In 1958, mathematician John von Neumann first used the term *singularity* to mean much the same as that meant by Kurzweil.

21. These articles debating the singularity appeared in a special issue of *IEEE Spectrum*, "The Singularity: Special Report," June 2008. Links to many of these articles and more (given that this debate has only begun) can be found on the *IEEE Spectrum* webpage, http://spectrum.ieee.org/biomedical/ethics/waiting-for-the-rapture.

22. Rodney Brooks, "I, Rodney Brooks, Am a Robot," *IEEE Spectrum*, June (2008): 71.

23. Jeff Hawkins and Sandra Blakeslee, *On Intelligence* (New York: Times Books, 2004), 87.

24. Ibid., 88.

25. Ibid.

26. McCullough, *Digital Ground*.

27. Hawkins and Blakeslee, *On Intelligence*.

28. Ibid.

29. These five phases of research are further elaborated in A. L. Threatt et al., "An Assistive Robotic Table for Older and Post-Stroke Adults: Results from Participatory Design and Evaluation Activities with Clinical Staff," in *Proceedings of CHI '14, the ACM Conference on Human Factors in Computing Systems* (Toronto: ACM Press, 2014), 673–682.

30. For an overview of personas and their use in related work, see A. Blomquist and M. Arvola, "Personas in Action: Ethnography in an Interaction Design Team," paper presented at the Proceedings of NordiCHI '02, the Second Nordic Conference on Human–Computer Interaction, Copenhagen, October 14–17, 2002.

31. J. Merino et al., "Forward Kinematic Model for Continuum Robotic Surfaces," in *2012 IEEE/RSJ International Conference on Intelligent Robots and Systems (IROS)* (New York: IEEE, 2012), 3453–3460. In preparing this section, I'm indebted to Jessica Merino for her thesis work in our lab as reported in her graduate thesis, J. Merino, *Continuum Robotic Surface: Forward Kinematics Analysis and Implementation* (Clemson University, 2013).

32. R. C. Loureiro et al., "Advances in Upper Limb Stroke Rehabilitation: A Technology Push,"*Medical & Biological Engineering & Computing* 49, no. 10 (2011): 1145–1156.

33. Ibid.

34. Ibid. See also J. Hidler and P. S. Lum, "The Road Ahead for Rehabilitation Robotics," *Journal of Rehabilitation Research and Development (JRRD)* 48, no. 4 (2011): vii–x.

35. Loureiro et al., "Advances in Upper Limb Stroke Rehabilitation," 1145–1156.

36. Ibid., and see H. I. Krebs et al., "A Paradigm Shift for Rehabilitation Robotics: Therapeutic Robots Enhance Clinician Productivity in Facilitating Patient Recovery,"*IEEE Engineering in Medicine and Biology Magazine* 27, no. 4 (2008): 61–70.

37. Gilles Deleuze and Félix Guattari, *Nomadology: The War Machine* (New York: Semiotext(e), 1986).

38. Ibid., 88, 51, 22, 45, 30 and 63 (the last, with italics in original).

39. Ibid., 18.

40. Ibid., 48.

41. Ibid., 50, 45, 22.

42. Ibid., 62.

43. Ibid., 96.

44. Castells introduced the concept of "space of flows" in Manuel Castells, *The Informational City: Information Technology, Economic Restructuring, and the Urban-Regional Process* (Cambridge, Mass.: B. Blackwell, 1989). With respect to specifically the considerations of this book, see also Manuel Castells, *The Rise of the Network Society. The Information Age: Economy, Society, and Culture* (Malden, Mass.: Wiley-Blackwell, 2010).

45. Harry Francis Mallgrave, *The Architect's Brain: Neuroscience, Creativity, and Architecture* (Chichester, UK: Wiley-Blackwell, 2010), 149.

46. This theory is elaborated in a canonical book for the discipline of architecture, Robert Venturi, *Complexity and Contradiction in Architecture* (New York: Museum of Modern Art, 1966). As considered in the epilogue of *Architectural Robotics*, Ignasi de Solá-Morales and his concept of a "Weak Architecture" is arguably a better reflection of vortical thinking than Venturi's dialectical *both/and* (see Ignasi de Solá-Morales, "Weak Architecture," in *Architecture Theory since 1968*, ed. C. Michal Hays (Cambridge, Mass.: MIT Press, 1998), 614–623.

47. Brooks, "I, Rodney Brooks, Am a Robot," 73. In this instance again, there is a marvelous analog in architectural thinking found in Joseph Rykwert, "Organic and Mechanical," *RES* (1992): 11–18. Here, Joseph Rykwert traces the commingling of the terms *organic* and *mechanical* in architectural thought, beginning with Vitruvius' treatise (where "the Latin *organicus* did not mean anything very different from *mechanicus*"), to Gottfried Semper, Owen Jones, and John Ruskin (all of whom cultivated "ideas about a new way of imitating nature, or relating the organism to the built form") (pp. 13 and 17). In the conclusion of his paper, Rykwert laments that while architects have long been preoccupied with nature as inspiration for decoration and form, there is not yet a "theory of architecture based on a direct appeal to ... the nature that biology and chemistry study"

(p.18). Perhaps some aspects of this book represent small steps in the direction Rykwert had anticipated.

48. Brooks, "I, Rodney Brooks, Am a Robot," 73–74.

49. Gilles Deleuze and Felix Guattari, "Concrete Rules and Abstract Machines,"*SubStance* 13, no. 3/4, issue 44–45 (1984): 7–19.

CHAPTER EIGHT

1. William J. Mitchell, *E-Topia: Urban Life, Jim—but Not as We Know It* (Cambridge, Mass.: MIT Press, 1999), 59.

2. Kevin Kelly, *Out of Control: The Rise of Neo-Biological Civilization* (Reading, Mass.: Addison-Wesley, 1994), 472 and 169.

3. Steven Levy, *Artificial Life: A Report from the Frontier Where Computers Meet Biology* (New York: Vintage Books, 1993), 8.

CHAPTER NINE

1. Maurice Sendak, *Where the Wild Things Are* (New York: HarperCollins, 1991).

2. J. K. Huysmans, *Against the Grain (A rebours)* (New York: Dover Publications, 1969), 2.

3. Ibid., 61.

4. Ibid., 62.

5. Ibid.

6. André Breton, *Mad Love (L'amour fou)*, A French Modernist Library (Lincoln: University of Nebraska Press, 1987), 15.

7. Guillaume de Lorris and Jean de Meun, *Le Roman de la rose* (Paris: H. Champion, 1982), 20896–20907. The many implications of the broader Pygmalion theme, of a statue's coming to life, is found in Kenneth Gross, *The Dream of the Moving Statue* (Ithaca, N.Y.: Cornell University Press, 1992).

8. Filarete and John Richard Spencer, *Treatise on Architecture (Being the Treatise by Antonio Di Piero Averlino, Known as Filarete)* (New Haven, Conn.: Yale University Press, 1965), book XXIV, 184r and v:315.

9. Carlo Mollino, *Il Messaggio Dalla Camera Oscura* ["The Message from the Dark Room"] (Turin: Chiantore, 1949), 76. On the same page of this work, the expansive-minded Mollino added the following revelations about the wondrous realm of possibilities:

> Conceived by way of an analogous procedure, Joseph Conrad records the inexorable and "natural" world of typhoons, tropical forests, and his characters: that is, *an assemblage of fragments*: A "character"—whether this be a man or something else from nature—known in reality as a finite number of physical, psychological and atmospheric fragments, is avowed and detracted with each subsequent investigation, reborn in poetic autonomy from inchoate elements assembled from selective intuition. Almost in spite of the intentions of the author and reader, this dynamic poetic movement is nearly an ineluctable creation that generates a plot of successive events: the common destiny of the author and reader. An entire world is automatically created in the interior of the

author from these fragments—angles of vision selected by the author that assist in the unwinding of that original intuition that had first selected them and which, subsequently, make a fantastic coherence.

10. Karsten Harries, "The Ethical Significance of Environmental Beauty," in *Architecture, Ethics, and the Personhood of Place*, ed. Gregory Caicco (Hanover, N.H.: University Press of New England, 2007), 137.

11. Henri Lefebvre, *The Production of Space*, trans. Donald Nicholson-Smith (Cambridge, Mass.: Blackwell, 1991), 36. In this canonical treatment of space, Lefebvre's analysis follows from a three-part dialectic between everyday practices and the perception of space (*le perçu*), representations or theories of space (*le conçu*) and directly "lived" space (*le vécu*).

12. Ibid., 313.

13. Ibid., 320.

14. Ibid., 308–309.

15. Ibid., 356.

CHAPTER TEN

1. H. Ishii and B. Ullmer, "Tangible Bits: Towards Seamless Interfaces between People, Bits and Atoms," in *Proceedings of CHI 1997, the ACM Conference on Human Factors in Computing Systems* (New York: ACM: 1997), 234–241.

2. Paul Dourish, *Where the Action Is: The Foundations of Embodied Interaction* (Cambridge, Mass.: MIT Press, 2001); and Alissa N. Antle, "Embodied Child Computer Interaction: Why Embodiment Matters," *Interactions* 16, 2 (March/April 2009): 27–30.

3. Antle, "Embodied Child Computer Interaction."

4. Ibid.

5. C. Rinaldi, *In Dialogue with Reggio Emilia* (London: Routledge, 2005).

6. Michio Kaku, *Physics of the Future: How Science Will Shape Human Destiny and Our Daily Lives by the Year 2100*, 1st ed. (New York: Doubleday, 2011), 16.

7. Ibid., 17.

8. Ibid., 15.

9. Ibid., 17.

10. UNESCO, "The Global Literacy Challenge." Available at unesdoc.unesco.org/images/0016/001631/163170e.pdf.

11. The National Center for Education Statistics (NCES), "The National Assessment of Adult Literacy (NAAL)." Available at http://www.independentmail.com/news/education/illiteracy-rates-among-adults-falls-south-carolina/. The NAAL is the most current and continual measure of adult literacy in the United States.

12. Henri Lefebvre, *The Production of Space*, trans. Donald Nicholson-Smith (Cambridge, Mass.: Blackwell, 1991), 238.

13. Nicholas Negroponte, *Soft Architecture Machines* (Cambridge, Mass.: MIT Press, 1975).
14. Hyperbody Research Group, TU Delft, "Muscle Body." Available at www.bk.tudelft.nl/en/about-faculty/departments/architectural-engineering-and-technology/organisation/hyperbody/research/applied-research-projects/muscle-body/.
15. Benedict Carey, *How We Learn: The Surprising Truth About When, Where, and Why It Happens* (New York: Random House, 2014).
16. Benedict Carey quoted in Tara Parker-Pope, "Better Ways to Learn." Available at http://well.blogs.nytimes.com/2014/10/06/better-ways-to-learn/?_php=true&_type=blogs&_r=1/.
17. James Paul Gee, *What Video Games Have to Teach Us About Learning and Literacy* (New York: Palgrave Macmillan, 2007).
18. L. S. Vygotsky, "Imagination and Creativity in the Adolescent," in *The Collected Works of L. S. Vygotsky*, ed. R. W. Rieber (New York: Plenum, 1998). For an example of how Vygotsky's theories inform the design of educations environments, see F. Decortis and L. Lentini, "A Socio-Cultural Perspective of Creativity for the Design of Educational Environments," *eLearning Papers* 1, no. 13 (2009) 1–10.
19. "Words become worlds" is the eloquent phrase offered by my co-investigator, Dr. Susan King Fullerton, Clemson University.
20. Parish-Morris cites five studies documenting that "dialogic reading with traditional books is important for children's emergent literary skills" (Julia Parish-Morris et al., "Once Upon a Time: Parent–Child Dialogue and Storybook Reading in the Electronic Era," *Mind, Brain & Education* 7, no. 3 [2013]: 206).
21. For an overview of the read-aloud, see, for instance, S. J. Barrentine, "Engaging with Reading through Interactive Read-Alouds,"*Reading Teacher* 50, no. 1 (1996): 36–43.
22. Louise Rosenblatt, *The Reader, the Text, the Poem: The Transactional Theory of the Literary Work* (Carbondale: Southern Illinois University Press, 1978).
23. K. Lance and R. Mark, "The Link between Public Libraries and Early Reading Success." Available at www.slj.com/2008/09/research/the-link-between-public-libraries-and-early-reading-success/#_.
24. On the library's mission to promote literacy, see, for example, H. H. Lyman, *Literacy and the Nation's Libraries* (Chicago: American Library Association, 1977).
25. C. Snow, M. S. Burns, and P. Griffin, eds., *Preventing Reading Difficulties in Young Children* (Washington, D.C.: National Academy Press, ERIC Document Reproduction Service No. ED 416 465, 1998).
26. There is a lot of evidence for the impact of read-alouds on childhood literacy. See, for instance: A. Wiseman, "Interactive Read Alouds: Teachers and Students Constructing Knowledge and Literacy Together," *Early Childhood Education Journal*, no. 38 (2011): 431–438; L. M. Justice and J. Kaderavek, "Using Shared Storybook Reading to Promote Emergent Literacy," *TEACHING Exceptional Children* 34, no. 4 (2002): 8–13; and W. H. Teale, "Reading Aloud to Young Children as a Classroom Instructional Activity: Insights from Research and Practice," in *On Reading Books to Children: Parents and Teachers*, ed. A. van Kleeck, S. A. Stahl, and E. B. Bauer (Mahwah, N.J.: Erlbaum Associates, 2003), 114–139.

27. See, for instance, H. K. Ezell and L. M. Justice, "Increasing the Print Focus of Adult-Child Shared Book Reading through Observational Learning," *American Journal of Speech-Language Pathology*, 9, no. 1 (2000): 36–47.

28. There is a lot of evidence for the positive impact of read-alouds on, specifically, language growth and school achievement. See, for instance, C. Chomsky, "Stages in Language Development and Reading Exposure," *Harvard Educational Review*, 42, no. 1 (1972): 1–33.

29. M. Stephens, "Tame the Web: Libraries and Technology." Available at http://tametheweb.com.

30. Ishii and Ullmer, "Tangible Bits," 234–251.

31. See A. Clark, *Being There: Putting Brain, Body and World Together Again* (Cambridge, Mass.: MIT Press, 1997); and J. P. Gee, *Situated Language and Learning: A Critique of Traditional Schooling* (London: Routledge, 2004).

32. See, for instance, B. Beastall, "Enchanting a Disenchanted Child: Revolutionising the Means of Education Using Information and Communication Technology and E-Learning," *British Journal of Sociology of Education*, 27, no. 1 (2006): 97–110.

33. For an interesting case using print and digital means, see M. Back et al., "Listen Reader: An Electronically Augmented Paper-Based Book," in *Proceedings of CHI '01, the ACM Conference on Human Factors in Computing Systems* (New York: ACM ,2001), 23–29.

34. For an interesting case, see F. Garzotto and M. Forfori, "Fate2: Storytelling Edutainment Experiences in 2d and 3d Collaborative Spaces," in *Proceedings of IDC 2006, the ACM Conference on Interaction Design and Children* (New York: ACM, 2006),113–116.

35. For an interesting case, see H. Alborni et al., "Designing Storyrooms: Interactive Storytelling Spaces for Children," in *Proceedings of DIS 2000, the ACM Conference on Designing Interactive Systems* (New York: ACM, 2000), 95–104.

36. M. Back et al., "Listen Reader."

37. S. Uras, D. Ardu, and M. Deriu, "Do Not Judge an Interactive Book by Its Cover: A Field Research," in *Proceedings of the 2012 Conference on Advances in Mobile Computing and Multimedia* (New York: ACM, 2012), 17–20.

38. M. Umaschi and J. Cassell, "Soft Toys with Computer Hearts: Building Personal Storytelling Environments," in *Proceedings of CHI 1997, the ACM Conference on Human Factors in Computing Systems, Extended Abstracts* (New York: ACM, 1997), 20–21.

39. A. Bobick et al., "The Kidsroom: A Perceptually-Based Interactive and Immersive Story Environment," *Presence: Teleoperators and Virtual Environments* 8, no. 4 (2000): 367–391; H. Alborni et al., "Designing Storyrooms."

40. J. Cassell and K. Ryokai, "Making Space for Voice: Technologies to Support Children's Fantasy and Storytelling," *Personal and Ubiquitous Computing*, no. 5, issue 3 (2001): 169–190.

41. See C. N. Jensen, W. Burleson, and J. Sadauskas, "Fostering Early Literacy Skills in Children's Libraries: Opportunities for Embodied Cognition and Tangible Technologies," in *Proceedings of IDC 2012, the ACM Conference on Interaction Design and Children* (New York: ACM, 2012), 50–59.

42. L. R. Sipe, *Storytime: Young Children's Literary Understanding in the Classroom* (New York: Teachers College Press, 2008).

43. See Wiseman, "Interactive Read Alouds," 431–438; Justice and Kaderavek, "Using Shared Storybook Reading to Promote Emergent Literacy"; and Teale, "Reading Aloud to Young Children as a Classroom Instructional Activity."

44. A. L. Baer, "Constructing Meaning through Visual Spatial Activities," *The Alan Review* 16, no. 2 (2009): 27–30.

45. This process of representation through some other system of signs is defined in the literature as *transmediation* (K. G. Short, G. Kauffman, and L. H. Kahn, "'I Just Need to Draw': Responding to Literature across Multiple Sign Systems," *Reading Teacher* 54, no. 2 [2000]: 160–171).

46. Shane Evans, *Underground* (New York: Roaring Brook Press, 2011).

47. The early version of proprietary Sifteo cubes we used are motion-aware, 1.5-inch blocks with full-color clickable screens that interact with users and each other when they are shaken, tilted, rotated, and placed adjacent to one another. They are designed for use by players ages 6 and up. We are quite easily able to hack the Sifteos for our purposes, adding our own images for the small screens as well as our own programming, as suggested in our descriptions, here. For more about the current version of Sifteo cubes, see https://www.sifteo.com/cubes.

48. Arduino boards are a popular hobbyist hardware and software platform for prototyping systems like this one. For more on the Arduino platform, see http://www.arduino.cc/.

49. J. Read and P. Markopoulos, "C15: Evaluating Children's Interactive Technology," Course notes distributed at CHI '11, the ACM Conference on Human Factors in Computing Systems, Vancouver, BC, May 7–12, 2011. These materials are elaborated in book form as P. Markopoulos, *Evaluating Children's Interactive Products: Principles and Practices for Interaction Designers* (Boston: Morgan Kaufmann, 2008).

50. See Read and Markopoulos, "C15: Evaluating Children's Interactive Technology."

51. I draw these conclusions from observation during this and other studies with children, as well as from the course discussion and content of the CHI course (Read and Markopoulos, "C15: Evaluating Children's Interactive Technology").

52. The value of qualitative outcomes of the kinds of research activities considered in this section may be contested, but this author and many peers would argue that such outcomes are part of a larger design consideration that should include design research precedents, theory, creative and formal design explorations, sustainability, and other inputs. We have returned, here, to the issue of the "wicked problem" considered in chapter 2 of this book.

53. Anne F. Rockwell and Megan Halsey, *Four Seasons Make a Year* (New York: Walker, 2004).

54. See George J. Schafer et al., "An Interactive, Cyber-Physical Read-Aloud Environment: Results and Lessons from an Evaluation Activity with Children and Their Teachers," in

Proceedings of DIS 2014, the ACM Conference on the Design of Interactive Systems (New York: ACM, 2014), 865–874. Much of this section draws from this particular paper.

55. On documenting and analyzing children's response initiations, see Sipe, *Storytime*.

56. On Baer's approach, see Baer, "Constructing Meaning through Visual Spatial Activities."

57. Sipe, *Storytime*.

58. L. B. Gambrell and R. J. Bales, "Mental Imagery and the Comprehension-Monitoring Performance of Fourth-and-Fifth-Grade Poor Readers," *Reading Research Quarterly*, no. 21 (1986): 454–464.

59. See a review of this research in D. Quenqua, "Quality of Words, Not Quantity, Is Crucial to Language Skills, Study Finds," *New York Times* (October 17, 2014): A22.

60. Michael S. A. Graziano, *Consciousness and the Social Brain* (Oxford: Oxford University Press, 2013).

61. Michael S. A. Graziano, "Are We Really Conscious?" *New York Times* (October 10, 2014), SR12.

62. Antle, "Embodied Child Computer Interaction."

63. Ferris Jabr, "The Reading Brain in the Digital Age: The Science of Paper Versus Screens." Available at www.scientificamerican.com/article/reading-paper-screens/.

64. Ibid.

65. There may not yet be longitudinal studies on the effects of screen-based devices on childhood learning and literacy, but there is much interest from the research community. For an overview of what research has been accomplished, the key issues, and where this research may lead, see D. Quenqua, "Is E-Reading to Your Toddler Story Time, or Simply Screen Time?" *New York Times* (October 11, 2014), A1.

66. Parish-Morris et al., "Once Upon a Time."

67. Jabr, "The Reading Brain in the Digital Age."

68. Ibid.

69. Ibid.

70. U.S. Department of Education, "National Education Technology Plan: Skills, Education & Competitiveness." Available at www.p21.org/storage/documents/21st_century_skills_education_and_competitiveness_guide.pdf; Partnership for 21st Century Skills, "Reimagining Citizenship for the 21st Century: A Call to Action for Policymakers and Educators." Available at www.p21.org/storage/documents/Reimagining_Citizenship_for_21st_Century_webversion.pdf; and the International Society for Technology in Education, "ISTE Standards: Students." Available at www.iste.org/standards/standards-for-library users.

71. Harry Francis Mallgrave, *The Architect's Brain: Neuroscience, Creativity, and Architecture* (Chichester, UK: Wiley-Blackwell, 2010), 210.

CHAPTER ELEVEN

1. Nicholas Negroponte, *Soft Architecture Machines* (Cambridge, Mass.: MIT Press, 1975).

CHAPTER TWELVE

1. I'm thankful to Mark D. Gross, director of the ATLAS Institute of the University of Colorado, Boulder, for suggesting this strategy of "What it's not" for defining architectural robotics.
2. Recall the definition of cyber-physical systems offered by Horváth and Gerritsen in chapter 1, note 9: "In the broadest sense, cyber-physical systems (CPSs) blend the knowledge and technologies of the third wave of information processing, communication and computing with the knowledge and technologies of physical artifacts and engineered systems."
3. U.S. National Science Foundation, "Cyber-Human Systems." Available at www.nsf.gov/cise/iis/chs_pgm13.jsp.
4. For more on the Fun Palace, see S. Matthews, *From Agit-Prop to Free Space: The Architecture of Cedric Price* (New York: Black Dog Publishing, 2007).
5. For more on the Plug-in-City, see Simon Sadler, *Archigram: Architecture without Architecture* (Cambridge, Mass.: MIT Press, 2005).
6. Pontus Hulten, "The Museum as a Communications Center," in *A New Cultural Center in Paris: Le Centre National D'art Et De Culture Georges Pompidou (Plateau Beaubourg 75004 Paris - Centre Pompidou)*, 9. Available at https://www.centrepompidou.fr/media/imgcoll/Collection/DOC/M5050/M5050_A/M5050_ARCV001_DP-1999015.pdf.
7. Hulten, "The Museum as a Communications Center," 9.
8. Ibid., 5.
9. More precisely, Andy Sedgwick is a high-ranked "fellow" employee of Arup, the engineering firm that (as *Ove Arup*) had partnered with Piano and Rogers in the realization of the *Centre Pompidou*.
10. Andy Sedgwick, "Would the Pompidou Get Built Today?" Available at http://thoughts.arup.com/post/details/290/would-the-pompidou-get-built-today.
11. The troubled state of Paris as a cultural capital was considered in Dan Bilefsky and Doreen Carvajal, "A Capital of the Arts Is Forced to Evolve," *New York Times* (October 28, 2014): A6.
12. The *Fondation Jérôme Seydoux-Pathé* is dedicated to preserving the history of French film company Pathé and promoting cinematography.
13. Robin Pogrebin, "British Architect Wins 2007 Pritzker Prize," *New York Times* (March 28, 2007). Available at http://www.nytimes.com/2007/03/28/arts/design/28cnd pritzker.html?pagewanted=2.
14. Robert McCarter, *Building; Machines* (New York: Pamphlet Architecture/Princeton Architectural Press, 1987), 10.
15. Walter Benjamin, "The Work of Art in the Age of Mechanical Reproduction," in *Illuminations*, ed. Hannah Arendt (New York: Harcourt, Brace & World, 1968), 239.

16. Ibid., 234.
17. Architecture is "appropriated in a twofold manner: by use and by perception—or rather, by touch and by sight." Ibid., 240.
18. Ibid., 240.
19. Stan Allen, "Dazed and Confused," *Assemblage* 27 (1995): 48.
20. Ibid., 50.
21. Ibid.
22. Ibid.
23. N. Katherine Hayles, "A New Framework for Meaning: The Cognitive Nonconscious," paper presented at the conference Critical and Clinical Cartographies, Delft, Netherlands, November 13–14, 2014. N. Katherine Hayles offered this definition for my ART/ *home+* project as a comment following my own talk at this conference, and then she elaborated the same the following day in her own talk.
24. N. Katherine Hayles, *How We Became Posthuman: Virtual Bodies in Cybernetics, Literature, and Informatics* (Chicago: University of Chicago Press, 1999), 291.
25. The Chinese room thought experiment was introduced in John Searle, "Minds, Brains and Programs," *Behavioral and Brain Sciences* 3, no. 3 (1980): 417–457.
26. The Turing test is a test of a machine's ability to exhibit intelligent behavior equivalent to, or indistinguishable from, that of a human ("The Turing Test, 1950," available at www.turing.org.uk/scrapbook/test.html).
27. Edwin Hutchins, "Distributed Cognition," 7. Available at http://gnowledge.org/~sanjay/ Advanced_Cogsci_Course/Week4/Week3_DistributedCognition_1.pdf.
28. Moravec is quoted from Daniel Crevier, *AI: The Tumultuous Search for Artificial Intelligence* (New York: Basic Books, 1993).
29. Hayles, *How We Became Posthuman*, 290.
30. I am drawing the quoted fragments from two sources: Hayles, *How We Became Posthuman*, 290; and "An Interview/Dialogue with Albert Borgmann and N. Katherine Hayles on Humans and Machines," 4. Available at www.press.uchicago.edu/Misc/ Chicago/borghayl.html.
31. Hayles, *How We Became Posthuman*, 289–290.
32. Ibid., 287.
33. Ibid.
34. Ibid.
35. Ibid., 290.
36. Likewise, while two forms of blue may not go together very well in your wardrobe, many different blues exhibited in a restaurant décor (on its walls, doors, trim, place settings, chairs, tables) provide us an opportunity to discover an accord between them.
37. "Cloud computing is computing in which large groups of remote servers are networked to allow centralized data storage and online access to computer services or resources"

(http://en.wikipedia.org/wiki/Cloud_computing). While cloud computing can be public, private, or a hybrid of the two, I am in these paragraphs referring specifically to the public version. Essentially, what I am promoting in these pages is a form of localized, community-based cloud computing that can connect with the other varieties or not.

38. This assessment and the bullets that follow are drawn from Zeynep Tufekci and Brayden King, "We Can't Trust Uber," *New York Times* (December 8, 2014): A27. The "Internet of Things" is defined and addressed later in this chapter.

39. Ibid.

40. The perspectives of Hawking and Musk are drawn from Nick Bilton, "Artificial Intelligence as a Threat," *New York Times* (November 6, 2014): E2.

41. The words of Goethe as quoted in Ignasi de Solá-Morales, "Weak Architecture," in *Architecture Theory since 1968*, ed. C. Michal Hays (Cambridge, Mass.: MIT Press, 1998), 621. Solá-Morales draws the concept of weak architecture from philosopher Gianni Vattimo's concept of *pensiero debole* ("weak thought") in Gianni Vattimo, Pier Aldo Rovatti, and Peter Carravetta, *Weak Thought*, SUNY Series in Contemporary Italian Philosophy (Albany: State University of New York Press, 2012).

42. I'm evoking the concept of the *form or formlessness* elaborated in Manfredo Tafuri, *Architecture and Utopia: Design and Capitalist Development* (Cambridge, Mass.: MIT Press, 1976).

43. For more on Apple's decision to entrust the design of its mobile operating system to Ive, its product designer extraordinaire, rather than to its software engineers, see Nick Wingfield and Nick Bilton, "Apple Shake-up Could Lead to Design Shift," *New York Times* (November 1, 2012): B1.

44. Solá-Morales, "Weak Architecture."

45. Robert Kaltenbrunner, "East of Planning: Urbanistic Strategies Today," *Daidalo* 72 (1999): 8.

46. "Urban Acupuncture." Available at http://en.wikipedia.org/wiki/Urban_acupuncture.

47. Erik Swyngedouw, "Metabolic Urbanization: The Making of Cyborg Cities," in *In the Nature of Cities: Urban Political Ecology and the Politics of Urban Metabolism*, ed. Nik Heynen, Maria Kaika, and E. Swyngedouw (New York: Routledge, 2006), quotations from, respectively, 35, 37, and 37 again.

48. Imre Horváth, "What the Design Theory of Social-Cyber-Physical Systems Must Describe, Explain and Predict?," in *An Anthology of Theories and Models of Design: Philosophy, Approaches and Empirical Explorations*, ed. Amaresh Chakrabarti and Lucienne T. M. Blessing (London: Springer, 2014), 116.

49. Ibid., 116.

50. See Malcolm McCullough, *Digital Ground: Architecture, Pervasive Computing, and Environmental Knowing* (Cambridge, Mass.: MIT Press, 2004). What is particularly compelling in *Digital Ground* is its insistence that we recognize and harness the capacity of familiar built environments (as mundane as the coffee shop, in one of McCullough's examples) to contribute in large part to the "programming" of a smart place.

51. Paul Kominers, "Interoperability Case Study: Internet of Things (IoT)," in the Berkman Center for Internet & Society Research Publication Series. The Berkman Center for Internet & Society. Available at http://papers.ssrn.com/sol3/papers.cfm?abstract_id=2046984.

52. Horváth, "What the Design Theory of Social-Cyber-Physical Systems Must Describe, Explain and Predict?," 106.

53. Ibid.

54. Imre Horváth, professor of industrial design engineering, identified an "Architectural Robotics future" with this third current of CPSs during a lengthy conversation we had together in his office at TU Delft on December 15, 2014.

55. Quoted from Walter Isaacson, *Steve Jobs* (New York: Simon & Schuster, 2011), 567.

56. "Artistically motivated" were the words used to describe Frank Gehry's exuberant Vuitton Foundation contemporary art center (Paris, 2014), commissioned by the wealthiest Frenchman, Bernard Arnault, chairman and chief executive of the luxury goods conglomerate LVMH, and inaugurated by none other than the nation's president, François Hollande (J. Giovannini, "An Architect's Big Parisian Moment," *New York Times* [October 20, 2014]: C1). As reported by the *Financial Times*, the motivation for the building was "to display the wealth and taste of the corporation and Bernard Arnault, its art-collecting owner and France's richest man" (Edwin Heathcote, "Frank Gehry's Fondation Louis Vuitton," available at http://www.ft.com/cms/s/0/4c8ad72e-5844-11e4-b331-00144feab7de.html).

57. These methods, in turn, are the established methods, or adaption of them, as developed within industrial engineering and human factors psychology.

58. Lewis Mumford, "The Corruption of Liberalism," *New Republic* (April 29, 1940): 569. It should be noted that Mumford is not against liberalism; he is instead distinguishing between what he refers to as "practical liberalism"" and "ideal liberalism."

59. Stewart Brand, "The Purpose of the Whole Earth Catalog," *Whole Earth Catalog* (Fall 1968). Issues of the *Whole Earth Catalog* are available online, beginning with this inaugural issue, at www.wholeearth.com/issue-electronic-edition.php?iss=1010.

60. Mumford, "The Corruption of Liberalism," 570.

61. An alarming statistic and a disparaging forecast for the future, as reported in David Brooks, "Our Machine Masters," *New York Times* (October 30, 2014): A31. The statistic, first: in 2001 the top 10 websites accounted for 31 percent of all U.S. page views, but by 2010, they accounted for 75 percent of them. This is one indicator of how a company that enters this "virtuous cycle," offers Kevin Kelly in the same article, can "grow so big, so fast, that it overwhelms any upstart competitors. As a result, our A.I. future is likely to be ruled by an oligarchy of two or three large, general-purpose cloud-based commercial intelligences." An outcome, offers Brooks, is that "people become less idiosyncratic, … everybody follows the prompts and chooses to be like each other."

62. René Descartes, "Discourse on the Method," part IV, 15. Available at www.earlymoderntexts.com/pdfs/descartes1637.pdf.

63. See chapter 1 of Hilary Putnam, *Reason, Truth, and History* (Cambridge: Cambridge University Press, 1981). Among works of fiction and films taking up the same theme is William Gibson's 1984 novel *Neuromancer*.
64. Harry Francis Mallgrave, *The Architect's Brain: Neuroscience, Creativity, and Architecture* (Chichester, UK: Wiley-Blackwell, 2010), 211.
65. Brooks, "Our Machine Masters."
66. Ibid.
67. Marcus Pollio Vitruvius, *De Architectura (Ten Books)*, trans. F. Granger (Cambridge, Mass.: Harvard University Press), book I, ch. 1, 7.
68. See Steven Levy, *Artificial Life: A Report from the Frontier Where Computers Meet Biology* (New York: Vintage Books, 1993), 9.
69. Claims of Paro as a therapeutic robot are taken from the manufacturer's webpage, "Paro Therapeutic Robot," available at www.parorobots.com/. Turkle speaks passionately about this incident in her *Ted Talk* of March 2012 (available at www.ted.com/talks/sherry_turkle_alone_together?language=en#t-12129), as well as in her book (referenced in chapter 1), Sherry Turkle, *Alone Together: Why We Expect More from Technology and Less from Each Other* (New York: Basic Books, 2011).
70. Illah Reza Nourbakhsh, *Robot Futures* (Cambridge, Mass.: MIT Press, 2013). From the lay press, the ethics surrounding "self-replicating nanobots," "armies of semi-intelligent robots," and other breeds of intelligent machines is considered, for example, by Bilton, "Artificial Intelligence as a Threat," E2.
71. A. Kapadia et al., "A Novel Approach to Rethinking the Machines in Which We Live: A Multidisciplinary Course in Architectural Robotics," *IEEE Robotics and Automation* 21, no. 3 (2014): 143–150.
72. As a designer of furniture and household objects, Mendini is not alone in this tendency to draw on familiar codes; designers other than Mendini but also designing for Alessi—famously, Philippe Starck—have drawn on this same resource.
73. I am here drawing on the theory of social semiotics as elaborated in R. Hodge and G. Kress, *Social Semiotics* (New York: Cornell University Press, 1988). According to the theory of social semiotics, the meaning of an object is formed through the associations it elicits for the viewer, given his or her interpretation, which follows in turn from his or her cultural understanding as well as the context in which the viewer engages the object. The theory assumes that there is no stability in "signs" or "codes." A related thesis can be found in Arthur C. Danto, "The Transfiguration of the Commonplace," *Journal of Aesthetics and Art Criticism* 33, no. 2 (1974): 139–148, which was expanded into the book of the same title published by Harvard University Press in 1981.
74. Alessandro and Francesco Mendini, "Atelier Mendini." Available at www.ateliermendini.it/index.php?page=69.
75. Henri Focillon, *The Life of Forms in Art* (New York: Zone Books, 1989), 149.
76. These two works, surely, have in turn inspired more recent ones, like Renzo Piano's *Cities for a Small Planet*, Ken Yeang's *Designing with Nature: The Ecological Basis for Architectural Design*, and *Cradle to Cradle* by Braungart and McDonough.

77. Madeira Interactive Technologies Institute and Carnegie Mellon University, "AHA: Augmented Human Assistance." Available at www.cmuportugal.org/tiercontent.aspx?id=5227.
78. I'm borrowing these words from Mallgrave, *The Architect's Brain*, 5.
79. IEEE (Institute of Electrical and Electronics Engineers), "The Internet of Things." Available at http://iot.ieee.org/about.html.
80. Janna Anderson and Lee Rainie, "The Internet of Things Will Thrive by 2025." Pew Research Internet Project (Washington, D.C.: Pew Research Center, 2014). Available at http://www.pewinternet.org/2014/05/14/internet-of-things/.
81. "Cisco Visual Networking Index: Global Mobile Data Traffic Forecast Update, 2013–2018." Available at www.cisco.com/c/en/us/solutions/collateral/service-provider/visual-networking-index-vni/white_paper_c11-520862.html.
82. Net neutrality means, first, that the prices charged by broadband companies will not be regulated (they can charge you what they want for providing Internet services); and second, that broadband lines would be neutral in delivering content to your computer (Internet providers cannot favor the delivery of content related to their own business interests—they have an interest in and/or the content of [mostly smaller] businesses that pay the Internet provider to deliver it favorably). The debate over Net neutrality rages on, as considered, for instance, by Farhad Manjoo, "In Net Neutrality Push, Internet Giants on the Sidelines," *New York Times* (November 11, 2014). The same issue of the *New York Times* contains two additional articles on Net neutrality.
83. Again, these obstacles are adroitly considered in Nourbakhsh, *Robot Futures* with respect to robotics increasingly becoming part of our everyday spaces.
84. On this point, both a compelling argument and a personal plea to architects comes from Malcolm McCullough. He offers, for one, that "Architectural elements ... accomplish half the work of tuning aggregations of portable and embedded technology" (McCullough, *Digital Ground*, 94).
85. Nourbakhsh, *Robot Futures*, 110. On the same page, Nourbakhsh identifies the current he is working against as the "empowerment of corporations" that manufacture and sell robotics and other ICT, which "can cause *disempowerment* in communities."
86. I'm invoking a digital interpretation of the "third place" described in Ramon Oldenburg and Dennis Brissett, "The Third Place," *Qualitative Sociology* 5 no. 4 (1982): 265–284.
87. I'm here combining fragments from two of Cage's outputs: the text from John Cage, *Mesostic for Elfriede Fischinger* (Los Angeles, Calif.: Center for Visual Music, Elfriede Fischinger Collection, 1980); and John Cage, *I-VI, The Charles Eliot Norton Lectures* (Cambridge, Mass.: Harvard University Press, 1990). Cage's delivery of his Norton Lecture is an astounding performance: seewww.ubu.com/sound/cage_norton.html.
88. Cage, *I-VI*.
89. Mark Wigley, "Recycling Recycling," in *Eco-Tec: Architecture of the in-Between* (New York: Princeton Architectural Press, 1999), 40.
90. David W. Orr, *The Nature of Design: Ecology, Culture, and Human Intention* (New York: Oxford University Press, 2002), 20.

91. The quoted fragment is from Orr, *The Nature of Design*, 3. Biophilia is "the urge to affiliate with other forms of life" (Edward O. Wilson, *Biophilia* [Cambridge, Mass.: Harvard University Press, 1984]).

92. These words are from an article appearing in *Wired* magazine's online content, written by Ilya Gelfenbeyn, cofounder and CEO of Speaktoit, a developer of human-computer interaction applications and platforms (Ilya Gelfenbeyn, "How New Technologies Return Us to Our Roots," available at www.wired.com/2013/08/how-new-technologies-return-us-to-our-roots/.) Here, Gelfenbeyn is specifically referring to a digitally embedded built environment, focusing specifically on the example of the kitchen and the new ways we can interact with it given current and near-future digital technologies. The paragraph from which these words are drawn is worthy of inclusion here:

> The developments I've been describing don't simply "enhance" our environment, or make it more "user-friendly." Rather, they completely alter the way we interact with it. We are in the midst of nothing short of a revolution in human perception.... I would argue, following scholars such as "the medium is the message" philosopher Marshall McLuhan, that this technology has the potential to return us to a more organic, integral way of being, to restore, in a way, the holistic world of the primitive man that earlier technologies dislocated us from.

BIBLIOGRAPHY

Alberti, Leon Battista. 1988. *On the Art of Building in Ten Books.* Trans. J. Rykwert, N. Leach and R. Tavernor. Cambridge, Mass.: MIT Press.

Alborni, H., A. Druin, J. Montemayor, M. Platner, J. Porteous, L. Sherman, A. Boltman, 2000. "Designing Storyrooms: Interactive Storytelling Spaces for Children." In *Proceedings of DIS 2000, the ACM Conference on Designing Interactive Systems*, 95–104. New York: ACM.

Alexander, Christopher, Sara Ishikawa, and Murray Silverstein. 1977. *A Pattern Language: Towns, Buildings, Construction.* New York: Oxford University Press.

Allen, Stan. 1995. "Dazed and Confused." *Assemblage* 27: 48.

Anderson, Janna, and Lee Rainie. 2014. "The Internet of Things Will Thrive by 2025." Pew Research Internet Project. Washington, D.C.: Pew Research Center.

"A New Cultural Center in Paris: Le Centre National D'art Et De Culture Georges Pompidou (Plateau Beaubourg 75004 Paris–Centre Pompidou)." Available at https://www.centre pompidou.fr/media/imgcoll/Collection/DOC/M5050/M5050_A/M5050_ARCV001_DP-1999015.pdf.

"An Interview/Dialogue with Albert Borgmann and N. Katherine Hayles on Humans and Machines." Available at http://www.press.uchicago.edu/Misc/Chicago/borghayl.html.

Antle, Alissa N. 2009. "Embodied Child Computer Interaction: Why Embodiment Matters." *Interactions* 16, no. 2 (March/April): 27–30.

Antón, Susan C., Richard Potts, and Leslie C. Aiello. 2014. "Evolution of Early Homo: An Integrated Biological Perspective." *Science* 345 (6192): 1236828.

Antonelli, Paola. 2001. *Workspheres: Design and Contemporary Work Styles.* New York: Museum of Modern Art and Harry N. Abrams.

Aristotle. 1943. *Generation of Animals.* Trans. A. L. Peck. Loeb Classical Library. Cambridge, Mass.: Harvard University Press.

Bachelard, Gaston. 1969. *The Poetics of Space.* Trans. M. Jolas. New York: Beacon Press.

Back, M., J. Cohen, R. Gold, S. Harrison, and S. Minneman. (2001). "Listen Reader: An Electronically Augmented Paper-Based Book." In *Proceedings of CHI '01, the ACM Conference on Human Factors in Computing Systems*, 23–29. New York: ACM.

Baecker, Ronald M. 1993. *Readings in Groupware and Computer Supported Cooperative Work.* San Mateo, Calif.: Morgan Kaufmann.

Baer, A. L. 2009. "Constructing Meaning through Visual Spatial Activities." *Alabama Review* 16 (2): 27–30.

Barrentine, S. J. 1996. "Engaging with Reading through Interactive Read-Alouds." *Reading Teacher* 50 (1): 36–43.

Beastall, B. 2006. "Enchanting a Disenchanted Child: Revolutionising the Means of Education Using Information and Communication Technology and E-Learning." *British Journal of Sociology of Education* 27, (1): 97–110.

Benjamin, Walter. 1968. "The Work of Art in the Age of Mechanical Reproduction." Trans. H. Zohn. In *Illuminations*, ed. Hannah Arendt, 217–252. New York: Harcourt, Brace & World.

Berman, S., and H. Stern. 2012. "Sensors for Gesture Recognition Systems." *IEEE Transactions on Systems, Man, and Cybernetics. Part C: Applications and Reviews* 42 (3): 277–290.

Bilefsky, Dan, and Doreen Carvajal. 2014 (October 28). "A Capital of the Arts Is Forced to Evolve." *New York Times*, A6.

Bilton, Nick. 2014 (November 6)."Artificial Intelligence as a Threat." *New York Times*, E2.

Blomquist, A., and M. Arvola. 2002. "Personas in Action: Ethnography in an Interaction Design Team." Paper presented at the proceedings of NordiCHI '02, the Second Nordic Conference on Human-Computer Interaction, Copenhagen, Denmark, October 14–17, 2002.

Bobick, A., S. Intille, J. Davis, F. Baird, C. Pinhanez, L. Campbell, Y. Ivanov, 2000. "The Kidsroom: A Perceptually-Based Interactive and Immersive Story Environment." *Presence* 8 (4): 367–391.

Bondarenko, O., and R. Janssen. 2005. "Documents at Hand: Learning from Paper to Improve Digital Technologies." In *Proceedings of CHI '05, the ACM Conference on Human Factors in Computing Systems*, 121–130. New York: ACM.

Borenstein, Greg. 2012. *Making Things See*. Sebastopol, Calif.: Maker Media.

Bork, Robert Odell, William W. Clark, and Abby McGehee. 2011. *New Approaches to Medieval Architecture. Avista Studies in the History of Medieval Technology, Science and Art*. Burlington, Vt.: Ashgate.

Borradori, Giovanna. 1988. Recoding Metaphysics: Strategies of Italian Contemporary Thought. In *Recoding Metaphysics: The New Italian Philosophy*, ed. Giovanna Borradori. Evanston, IL: Northwestern University Press.

Borradori, Giovanna. 1987/88. "'Weak Thought' and Post Modernism: The Italian Departure from Deconstruction." *Social Text* 18 (Winter): 39–49.

Brand, Stewart. "The Purpose of the Whole Earth Catalog." *Whole Earth Catalog*, Fall, 1968.

Brautigan, Richard. 1967. "All Watched over by Machines of Loving Grace." In *All Watched over by Machines of Loving Grace*. San Francisco, Calif.: Communication Company.

Breton, André. 1987. *Mad Love (L'amour Fou). A French Modernist Library*. Lincoln: University of Nebraska Press.

Brooks, David. 2014 (October 30). "Our Machine Masters." *New York Times*, A31.

Brooks, Rodney. 2008. "I, Rodney Brooks, Am a Robot." *IEEE Spectrum* 45, no. 6 (June): 71–75.

Cage, John. 1980. *Mesostic for Elfriede Fischinger*. Los Angeles, Calif.: Center for Visual Music, Elfriede Fischinger Collection.

Cage, John. 1990. *I-VI. The Charles Eliot Norton Lectures*. Cambridge, Mass.: Harvard University Press.

Card, Stuart K., Thomas P. Moran, and Allen Newell. 1983. *The Psychology of Human-Computer Interaction*. Hillsdale, N.J.: L. Erlbaum Associates.

Carey, Benedict. 2014. *How We Learn: The Surprising Truth About When, Where, and Why It Happens*. New York: Random House.

Carlisle, James H. "Evaluating the Impact of Office Automation on Top Management Communication." In *American Federation of Information Processing Societies. AFIPS Conference Proceedings. Volume 45. 1976 National Computer Conference, June 7–10, 1976, New York*, 611–616. Montvale, N.J.: AFIPS Press.

Cassell, J. 1998. "A Framework for Gesture Generation and Interpretation." In *Computer Vision in Human-Machine Interaction*, ed. R. Cipolla and A. Pentland, 191–216. Cambridge: Cambridge University Press.

Cassell, J., and K. Ryokai. 2001. "Making Space for Voice: Technologies to Support Children's Fantasy and Storytelling." *Personal and Ubiquitous Computing* 5: 169–190.

Castells, Manuel. 1989. *The Informational City: Information Technology, Economic Restructuring, and the Urban-Regional Process*. Cambridge, Mass.: B. Blackwell.

Castells, Manuel. 2010. *The Rise of the Network Society. The Information Age: Economy, Society, and Culture*. Malden, Mass.: Wiley-Blackwell.

Chomsky, C. 1972. "Stages in Language Development and Reading Exposure." *Harvard Educational Review* 42: 1–33.

Churchman, C. West. 1967. "Guest Editorial: Wicked Problems." *Management Science* 14 (4): 141–142.

"Cisco Visual Networking Index: Global Mobile Data Traffic Forecast Update, 2013–2018." Available at www.cisco.com/c/en/us/solutions/collateral/service-provider/visual-networking-index-vni/white_paper_c11-520862.html.

Clark, A. 1997. *Being There: Putting Brain, Body and World Together Again*. Cambridge, Mass.: MIT Press.

Coplien, James O. 2006. "Organizational Patterns." In *Enterprise Information Systems VI*, ed. Isabel Seruca, José Cordeiro, Slimane Hammoudi and Joaquim Filipe, 43–52. Rotterdam: Springer.

Crevier, Daniel. 1993. *AI: The Tumultuous Search for Artificial Intelligence*. New York: Basic Books.

Danto, Arthur C. 1974. "The Transfiguration of the Commonplace." *Journal of Aesthetics and Art Criticism* 33 (2): 139–148.

Danto, Arthur C. 1981. *The Transfiguration of the Commonplace*. Cambridge, Mass.: Harvard University Press.

Dautenhahn, K. 2007. "Socially Intelligent Robots: Dimensions of Human-Robot Interaction." *Philosophical Transactions of the Royal Society of London. Series B, Biological Sciences* 362 (1480): 679–704.

de Certeau, Michel. 1984. *The Practice of Everyday Life*. Berkeley: University of California Press.

de Chirico, Giorgio. 1992. *Hebdomeros*. Trans. J. Ashbery. Cambridge, Mass.: Exact Change.

de Chirico, Giorgio. 1992. "Statues, Furniture and Generals." Trans. John Ashbery. In *Hebdomeros*. Cambridge, Mass.: Exact Change.

Decortis, F., and L. Lentini. 2009. "A Socio-Cultural Perspective of Creativity for the Design of Educational Environments." *eLearning Papers* 1 (13): 1–9.

Deleuze, Gilles, and Felix Guattari. 1984. "Concrete Rules and Abstract Machines." *SubStance* 13, no. 3/4 (44–45): 7–19.

Deleuze, Gilles, and Félix Guattari. 1986. *Nomadology: The War Machine*. New York: Semiotext(e).

de Lorris, Guillaume, and Jean de Meun. 1982. *Le Roman De La Rose*. Paris: H. Champion.

Descartes, René. "Discourse on the Method." Available at www.earlymoderntexts.com/pdfs/descartes1637.pdf.

de Solá-Morales, Ignasi. 1998. "Weak Architecture." In *Architecture Theory since 1968*, ed. C. Michal Hays, 614–623. Cambridge, Mass.: MIT Press.

Dolwick, J. S. 2009. "'The Social' and Beyond: Introducing Actor-Network Theory." *Journal of Maritime Archaeology* 4 (1): 21–49.

Dourish, Paul. 2001. *Where the Action Is: The Foundations of Embodied Interaction*. Cambridge, Mass.: MIT Press.

Dow, S., J. Lee, C. Oezbek, B. MacIntyre, J. D. Bolter, and M. Gandy. 2005. "Wizard of Oz Interfaces for Mixed Reality Applications." In *Proceedings of CHI '05, the ACM Conference on Human Factors in Computing Systems, Extended Abstracts*, 1339–1342. New York: ACM.

Eco, Umberto. 1989. *The Open Work*. Trans. A. Cancogni. Cambridge, Mass.: Harvard University Press.

Edelman, Gerald M. 2006. *Second Nature: Brain Science and Human Knowledge*. New Haven: Yale University Press.

Eden, Ammon H., Eric Steinhart, David Pearce, and James H. Moor. 2012. "Singularity Hypotheses: An Overview." In *Singularity Hypotheses: A Scientific and Philosophical Assessment*, ed. James H. Moor, Ammon H. Eden, Johnny H. Soarker, and Eric Steinhart, 1–12. New York: Springer.

Evans, Shane. 2011. *Underground*. New York: Roaring Brook Press.

Ezell, H. K., and L. M. Justice. 2000. "Increasing the Print Focus of Adult-Child Shared Book Reading through Observational Learning." *American Journal of Speech-Language Pathology* 9: 36–47.

Filarete, and John Richard Spencer. 1965. *Treatise on Architecture (Being the Treatise by Antonio di Piero Averlino, Known as Filarete)*. New Haven, Conn.: Yale University Press.

Focillon, Henri. 1989. *The Life of Forms in Art*. New York: Zone Books.

Fong, T., I. Nourbakhsh, and K. Dautenhahn. 2003. "A Survey of Socially Interactive Robots." *Robotics and Autonomous Systems* 42: 143–166.

Frayling, C. 1993. "Research in Art and Design." *Royal College of Art Research Papers* 1 (1): 1–5.

Gambrell, L. B., and R. J. Bales. 1986. "Mental Imagery and the Comprehension-Monitoring Performance of Fourth-and-Fifth-Grade Poor Readers." *Reading Research Quarterly* 21: 454–464.

Gamma, Erich, Richard Helm, Ralph Johnson, and John Vlissides. 1995. *Design Patterns: Elements of Reusable Object-Oriented Software*. Reading, Mass.: Addison-Wesley.

Garzotto, F., and M. Forfori. 2006. "Fate2: Storytelling Edutainment Experiences in 2D and 3D Collaborative Spaces." In *Proceedings of IDC 2006, the ACM Conference on Interaction Design and Children*, 113–116. New York: ACM.

Gee, J. P. 2004. *Situated Language and Learning: A Critique of Traditional Schooling*. London: Routledge.

Gee, James Paul. 2007. *What Video Games Have to Teach Us About Learning and Literacy*. New York: Palgrave Macmillan.

Gelfenbeyn, Ilya. "How New Technologies Return Us to Our Roots." Available at www.wired.com/2013/08/how-new-technologies-return-us-to-our-roots/.

Gibson, James J. 1979. *The Ecological Approach to Visual Perception*. Boston: Houghton Mifflin.

Giedion, Sigfried. 1948. *Mechanization Takes Command, a Contribution to Anonymous History*. New York: Oxford University Press.

Giovannini, J. "An Architect's Big Parisian Moment." *The New York Times*, October 20, 2014.

Graziano, Michael S. A. 2013. *Consciousness and the Social Brain*. Oxford: Oxford University Press.

Graziano, Michael S. A. 2014 (October 10). "Are We Really Conscious?" *New York Times*.

Green, K. E., L. J. Gugerty, J. C. Witte, I. D. Walker, H. Houayek, J. Rubinstein, R. Daniels, 2008. "Configuring an 'Animated Work Environment': A User-Centered Design Approach." Paper presented at IE 2008: the 4th International Conference on Intelligent Environments, Seattle, Washington, July 21–22, 2008.

Grönvall, Erik, Sofie Kinch, Marianne Graves Petersen, and Majken K. Rasmussen. 2014. "Causing Commotion with a Shape-Changing Bench: Experiencing Shape-Changing Interfaces in Use." In *Proceedings of CHI '14, the ACM Conference on Human Factors in Computing Systems*, 2559–2568. New York: ACM.

Gross, Kenneth. 1992. *The Dream of the Moving Statue*. Ithaca, N.Y.: Cornell University Press.

Guizzo, Erico. 2014. "So, Where Are My Robot Servants?" *IEEE Spectrum* (June): 74–79.

Harries, Karsten. 2007. "The Ethical Significance of Environmental Beauty." In *Architecture, Ethics, and the Personhood of Place*, ed. Gregory Caicco, 134–150. Hanover, N.H.: University Press of New England.

Hawkins, Jeff, and Sandra Blakeslee. 2004. *On Intelligence*. New York: Times Books.

Hayles, Katherine. 1999. *How We Became Posthuman: Virtual Bodies in Cybernetics, Literature, and Informatics*. Chicago: University of Chicago Press.

Hayles, N. Katherine. 2014. "A New Framework for Meaning: The Cognitive Nonconscious." Paper presented at the Critical and Clinical Cartographies Conference, Delft, Netherlands, November 13–14, 2014.

Heathcote, Edwin. "Frank Gehry's *Fondation Louis Vuitton*." Available at http://www.ft.com/intl/cms/s/0/4c8ad72e-5844-11e4-b331-00144feab7de.html#axzz3fcNb6QKL/.

Henig, Robin Marantz. 2007. "The Real Transformers." *New York Times Magazine* 156 (54020): 28–55.

Hidler, J., and P. S. Lum. 2011. "The Road Ahead for Rehabilitation Robotics." *Journal of Rehabilitation Research and Development* 48 (4): vii–x.

Hodge, R., and G. Kress. 1988. *Social Semiotics*. New York: Cornell University Press.

Homer. 2007. *The Odyssey*. Trans. I. C. Johnston. Arlington, Va.: Richer Resources Publications.

"Hong Kong Space Saver." Available at www.youtube.com/watch?v=f-iFJ3ncIDo/.

Horváth, Imre. 2014. "What the Design Theory of Social-Cyber-Physical Systems Must Describe, Explain and Predict?" In *An Anthology of Theories and Models of Design: Philosophy, Approaches and Empirical Explorations*, ed. Amaresh Chakrabarti and Lucienne T. M. Blessing, 99–120. London: Springer.

Horváth, Imre, and Bart H. M. Gerritsen. 2012. "Cyber-Physical Systems: Concepts, Technologies and Implementation Principles." In *TMCE 2012, the International Symposium on Tools and Methods of Competitive Engineering*, ed. Z. Rusák, I. Horváth, A. Albers, and M. Behrendt, 19–36. Karlsruhe, Germany.

Houayek, Henrique, Keith Evan Green, Leo Gugerty, Ian D. Walker, and James Witte. 2014. "AWE: An Animated Work Environment for Working with Physical and Digital Tools and Artifacts." *Personal and Ubiquitous Computing* 18: 1227–1241.

Hulten, Pontus. "The Museum as a Communications Center." In *A New Cultural Center in Paris: Le Centre National D'art Et De Culture Georges Pompidou (Plateau Beaubourg 75004 Paris - Centre Pompidou)*, 9. Available at https://www.centrepompidou.fr/media/imgcoll/Collection/DOC/M5050/M5050_A/M5050_ARCV001_DP-1999015.pdf.

Hutchins, Edwin. "Distributed Cognition." Available at http://gnowledge.org/~sanjay/Advanced_Cogsci_Course/Week4/Week3_DistributedCognition_1.pdf.

Huysmans, J. K. 1969. *Against the Grain (A rebours)*. New York: Dover Publications.

Hyperbody Research Group, TU Delft. "Muscle Body." Available at www.bk.tudelft.nl/en/about-faculty/departments/architectural-engineering-and-technology/organisation/hyperbody/research/applied-research-projects/muscle-body/.

IEEE. "The Internet of Things." Available at http://iot.ieee.org/about.html.

IEEE. 2008. "The Singularity: Special Report." *IEEE Spectrum* (June).

Ihde, Don. *Embodied Technics*. 2010. Copenhagen: Automatic Press.

International Society for Technology in Education, "International Standards: Students." Available at www.iste.org/standards/standards-for-library users/.

Isaacson, Walter. 2011. *Steve Jobs*. New York: Simon & Schuster.

Ishii, H., and B. Ullmer. 1997. "Tangible Bits: Towards Seamless Interfaces between People, Bits and Atoms." In *Proceedings of CHI 1997, the ACM Conference on Human Factors in Computing Systems*, 234–251. New York: ACM.

Jabr, Ferris. 2013. "The Reading Brain in the Digital Age: The Science of Paper Versus Screens." Available at www.scientificamerican.com/article/reading-paper-screens/.

Jensen, C. N., W. Burleson, and J. Sadauskas. 2012. "Fostering Early Literacy Skills in Children's Libraries: Opportunities for Embodied Cognition and Tangible Technologies." In *Proceedings of IDC 2012, the ACM Conference on Interaction Design and Children*, 50–59. New York, ACM.

Johanson, B., A. Fox, and T. Winograd. 2002. "The Interactive Workspaces Project: Experiences with Ubiquitous Computing Rooms." *IEEE Pervasive Computing* 1 (2): 67–74.

Jordan, Brigitte. 2013. "Pattern Recognition in Human Evolution and Why It Matters for Ethnography, Anthropology, and Society." In *Advancing Ethnography in Corporate Environments: Challenges and Emerging Opportunities*, ed. Brigitte Jordan, 193–213. Walnut Creek, Calif.: Left Coast Press.

Justice, L. M., and J. Kaderavek. 2002. "Using Shared Storybook Reading to Promote Emergent Literacy." *Teaching Exceptional Children* 34 (4): 8–13.

Kahn, Peter H., Nathan G. Freier, Takayuki Kanda, Hiroshi Ishiguro, Jolina H. Ruckert, Rachel L. Severson, and Shaun K. Kane. 2008. "Design Patterns for Sociality in Human-Robot Interaction." In *Proceedings of the 3rd ACM/IEEE International Conference on Human-Robot Interaction*, 97–104. Amsterdam: ACM.

Kaku, Michio. 2011. *Physics of the Future: How Science Will Shape Human Destiny and Our Daily Lives by the Year 2100*. New York: Doubleday.

Kaltenbrunner, Robert. 1999. "East of Planning: Urbanistic Strategies Today." *Daidalo* 72: 9–7.

Kant, Immanuel. *Critique of Pure Reason*. Trans. J. M. D. Meiklejohn. Dover Philosophical Classics, ed. Jim Manis. Hazleton, Pa.: The Electronics Classic Series.

Kapadia, A., I. Walker, K. E. Green, J. Manganelli, H. Houayek, A. M. James, V. Kanuri, 2014. "A Novel Approach to Rethinking the Machines in Which We Live: A Multidisciplinary Course in Architectural Robotics." *IEEE Robotics and Automation* 21 (3): 143–150.

Kelly, Kevin. 1994. *Out of Control: The Rise of Neo-Biological Civilization*. Reading, Mass.: Addison-Wesley.

Kessler, Aaron M. 2014 (October 6). "Technology Takes the Wheel." *New York Times*, B1.

Kidd, A. 1994. "The Marks Are on the Knowledge Worker." In *Proceedings of CHI '94, the ACM Conference on Human Factors in Computing Systems*, 186–191. New York: ACM.

Kinch, Sofie, Erik Grönvall, Marianne Graves Petersen, and Majken Kirkegaard Rasmussen. 2014. "Encounters on a Shape-Changing Bench: Exploring Atmospheres and Social Behaviour in Situ." In *Proceedings from TEI '14, the International Conference on Tangible, Embedded and Embodied Interaction*, 233–240. New York: ACM.

Köhler, Wolfgang. 1992. *Gestalt Psychology: An Introduction to New Concepts in Modern Psychology*. New York: Liveright.

Komatsu, T. 2006. "Audio Subtle Expressions Affecting User's Perceptions." In *Proceedings of IUI 2006, the International Conference on Intelligent User Interface*, 306–308. New York: ACM.

Kominars, Paul. "Interoperability Case Study: Internet of Things (IoT)." The Berkman Center for Internet & Society Research Publication Series. The Berkman Center for Internet & Society. Available at http://papers.ssrn.com/sol3/papers.cfm?abstract_id=2046984/.

Kosko, Bart. 1988. "Bidirectional Associative Memories." *IEEE Transactions on Systems, Man, and Cybernetics* 18 (1): 49–60.

Kowka, M., J. Johnson, H. Houayek, I. Dunlop, I. D. Walker, and K. E. Green. 2008. "The AWE Wall: A Smart Reconfigurable Robotic Surface." Paper presented at IE 2008, the 4th International Conference on Intelligent Environments, Seattle, Washington, July 21–22, 2008.

Krebs, H. I., S. Levy-Tzedek, S. E. Fasoli, A. Rykman-Berland, J. Zipse, J. A. Fawcett, J. Stein, 2008. "A Paradigm Shift for Rehabilitation Robotics: Therapeutic Robots Enhance Clinician Productivity in Facilitating Patient Recovery." *IEEE Engineering in Medicine and Biology Magazine* 27 (4): 61–70.

Kreith, Frank, and D. Yogi Goswami. 2005. *The CRC Handbook of Mechanical Engineering. Mechanical Engineering Handbook Series*. Boca Raton, Fla.: CRC Press.

Kurzweil, Ray. 2005. *The Singularity Is Near: When Humans Transcend Biology*. New York: Viking.

Kwak, Matthijs, Kasper Hornbæk, Panos Markopoulos, and Miguel Bruns Alonso. 2014. "The Design Space of Shape-Changing Interfaces: A Repertory Grid Study." In *Proceedings of the CHI '14, the ACM Conference on Human Factors in Computing Systems*, 181–190. New York: ACM.

Lakoff, George, and Mark Johnson. 1999. *Philosophy in the Flesh: The Embodied Mind and Its Challenge to Western Thought*. New York: Basic Books.

Lance, K., and R. Mark. "The Link between Public Libraries and Early Reading Success." Available at www.slj.com/2008/09/research/the-link-between-public-libraries-and-early-reading-success/#/.

Lee, C. S. George. 1982. "Robot Arm Kinematics, Dynamics, and Control." *Computer* 15 (12): 62–80.

Lefebvre, Henri. 1991. *The Production of Space*. Trans. D. Nicholson-Smith. Cambridge, Mass.: Blackwell.

Lehrer, Jonah. 2012 (January 30). "Groupthink: The Brainstorming Myth." *New Yorker*, 22–27.

Lepore, Jill. 2014 (May 12). "Away from My Desk." *New Yorker*, 72–76.

Levy, Steven. 1993. *Artificial Life: A Report from the Frontier Where Computers Meet Biology*. New York: Vintage Books.

Levy, Steven. 2011 (January). "The A.I. Revolution." *Wired*, 88–89.

Lewis, Nigel. 1994. *The Book of Babel: Words and the Way We See Things*. Iowa City: University of Iowa Press.

Lifton, Robert Jay. 1994. *The Protean Self: Human Resilience in an Age of Fragmentation*. New York: Basic Books.

Linehan, Conor, Ben J. Kirman, Stuart Reeves, Mark A. Blythe, Joshua G. Tanenbaum, Audrey Desjardins, and Ron Wakkary. 2014. "Alternate Endings: Using Fiction to Explore Design Futures." In *CHI '14 Extended Abstracts on Human Factors in Computing Systems*, 45–48. Toronto: ACM.

Loureiro, R. C., W. S. Harwin, K. Nagai, and M. Johnson. 2011. "Advances in Upper Limb Stroke Rehabilitation: A Technology Push." *Medical & Biological Engineering & Computing* 49 (10): 1145–1156.

Luff, P., C. Heath, H. Kazuoka, K. Yamakazi, and J. Yamashita. 2006. "Handling Documents and Discriminating Objects in Hybrid Spaces." In *Proceedings of CHI '06, the ACM Conference on Human Factors in Computing Systems*, 561–570. New York: ACM.

Lyman, H. H. 1977. *Literacy and the Nation's Libraries*. Chicago: American Library Association.

Madeira Interactive Technologies Institute and Carnegie Mellon University. "AHA: Augmented Human Assistance." Available at www.cmuportugal.org/tiercontent.aspx?id=5227/.

Mallgrave, Harry Francis. 2010. *The Architect's Brain: Neuroscience, Creativity, and Architecture*. Chichester, UK: Wiley-Blackwell.

Malone, T. W. 1983. "How Do People Organize Their Desks? Implications for the Design of Office Information Systems." *ACM Transactions on Office Information Systems* 1 (1): 99–112.

Manjoo, Farhad. 2014 (November 11). "In Net Neutrality Push, Internet Giants on the Sidelines." *New York Times*, B1.

Marinetti, F. T. 1991. "'The Birth of a Futurist Aesthetic' from *War, the World's Only Hygiene*" [1911–1915]. In *Let's Murder the Moonshine: Selected Writings*, trans. R. W. Flint. Los Angeles: Sun and Moon.

Markoff, John. 2013 (May 21). "In 1949, He Imagined an Age of Robots." *New York Times*, D8.

Markoff, John. 2014 (September 1). "Brainy, Yes, but Far from Handy." *New York Times*, D1.

Markhoff, John, and Claire Cain Miller. 2014 (June 17). "Danger: Robots Working." *New York Times*, D1 and D5.

Markopoulos, P. 2008. *Evaluating Children's Interactive Products: Principles and Practices for Interaction Designers*. The Morgan Kaufmann Series in Interactive Technologies. Boston: Morgan Kaufmann.

Mataric, Maja J. 2007. *The Robotics Primer*. Cambridge, Mass.: MIT Press.

Matthews, S. 2007. *From Agit-Prop to Free Space: The Architecture of Cedric Price*. New York: Black Dog Publishing.

McCarter, Robert. 1987. *Building; Machines*. New York: Pamphlet Architecture/Princeton Architectural Press.

McCullough, Malcolm. 2004. *Digital Ground: Architecture, Pervasive Computing, and Environmental Knowing*. Cambridge, Mass.: MIT Press.

McEwen, Adrian, and Hakim Cassimally. 2014. *Designing the Internet of Things*. West Sussex, UK: Wiley.

McHale, John. 1969. *The Future of the Future*. New York: G. Braziller.

Mendini, Alessandro, and Francesco Mendini. "Atelier Mendini." Available at www.ateliermendini.it/index.php?page=69/.

Merino, J. 2013. *Continuum Robotic Surface: Forward Kinematics Analysis and Implementation*. MS diss., Clemson University.

Merino, J., A. L. Threatt, I. D. Walker, and K. E. Green. 2012. "Forward Kinematic Model for Continuum Robotic Surfaces." In *2012 IEEE/RSJ International Conference on Intelligent Robots and Systems (IROS)*, 3453–3460. New York: IEEE.

Mitchell, William J. 1999. *E-Topia: Urban Life, Jim—but Not as We Know It*. Cambridge, Mass.: MIT Press.

Mitra, S., and T. Acharya. 2007. "Gesture Recognition: A Survey." *IEEE Transactions on Systems, Man, and Cybernetics. Part C, Applications and Reviews* 37 (3): 311–324.

Mollino, Carlo. 1949. "Utopia E Ambientazione." *Domus* 237 (August): 14–19.

Mollino, Carlo. 1949. *Il Messaggio Dalla Camera Oscura* ["The Message from the Dark Room"]. Turin: Chiantore.

Mumford, Lewis. 1940 (April 29). "The Corruption of Liberalism." *New Republic*, 568–573.

Mutlu, B., C. Bartneck, J. Ham, V. Evers, and T. Kanda. 2011. "Social Robotics." Paper presented at the proceedings of ICSR 2011, the Third International Conference on Social Robotics, Amsterdam, November 24–25, 2011.

National Center for Education Statistics (NCES). "The National Assessment of Adult Literacy (NAAL)." Available at http://www.independentmail.com/news/2009/jan/10/illiteracy-rates-among-adults-falls-south-carolina/.

Negroponte, Nicholas. 1975. *Soft Architecture Machines*. Cambridge, Mass.: MIT Press.

Neutra, Richard Joseph. 1954. *Survival through Design*. New York: Oxford University Press.

Nocks, Lisa. 2007. *The Robot: The Life Story of a Technology*. Westport, Conn.: Greenwood Press.

Norman, Donald A. 1988. *The Psychology of Everyday Things*. New York: Basic Books.

Norman, Donald A. 2007. "The Next UI Breakthrough, Part 2: Physicality." *Interactions* 14 issue 4 (July–August): 46–47.

Norman, Donald A. 2013. *The Design of Everyday Things*. New York: Basic Books.

Nourbakhsh, Illah Reza. 2013. *Robot Futures*. Cambridge, Mass.: MIT Press.

Oates, Phyllis Bennett. 1981. *The Story of Western Furniture*. London: Herbert Press.

Oldenburg, Ramon, and Dennis Brissett. 1982. "The Third Place." *Qualitative Sociology* 5 (4): 265–284.

Orr, David W. 2002. *The Nature of Design: Ecology, Culture, and Human Intention*. New York: Oxford University Press.

Ortega y Gasset, José. 1972 [1925]. "La Deshumanizacíon del arte e Ideas sobre la Novella." In *The Dehumanization of Art and Other Essays on Art, Culture and Literature*, trans. H. Weyl, 33. Princeton, N.J.: Princeton University Press.

"Pain in Non-Verbal Elderly Largely Undetected by Family Caregivers." Available at http://m2.facebook.com/americanpainsociety/.

Parish-Morris, Julia, Neha Mahajan, Kathy Hirsh-Pasek, Roberta Michnick Golinkoff, and Molly Fuller Collins. 2013. "Once Upon a Time: Parent–Child Dialogue and Storybook Reading in the Electronic Era." *Mind, Brain and Education: The Official Journal of the International Mind, Brain, and Education Society* 7 (3): 200–211.

Park, K., Z. Bien, J. Lee, K. Byung, J. Lim, J. Kim, 2007. "Robotic Smart House to Assist People with Movement Disabilities." *Autonomous Robots* 22: 183–198.

Parker-Pope, Tara. "Better Ways to Learn." Available at http://well.blogs.nytimes.com/2014/10/06/better-ways-to-learn/?_php=true&_type=blogs&_r=0/.

"PARO Therapeutic Robot." Available at www.parorobots.com/.

Partnership for 21st Century Skills. "Reimagining Citizenship for the 21st Century: A Call to Action for Policymakers and Educators." Available at www.p21.org/storage/documents/Reimagining_Citizenship_for_21st_Century_webversion.pdf.

Pask, Gordon. 1969. "The Architectural Relevance of Cybernetics." *Architectural Design* (September): 494–496.

Pi, Youguo, Wenzhi Liao, Mingyou Liu, and Jianping Lu. 2008. "Theory of Cognitive Pattern Recognition." In *Pattern Recognition Techniques, Technology and Applications*, ed. Peng-Yeng Yin, 433–462. Vienna: InTech.

Poe, Edgar Allan. 1960. *The Fall of the House of Usher, and Other Tales*. New York: New American Library.

Pogrebin, Robin. 2007 (March 28). "British Architect Wins 2007 Pritzker Prize." *New York Times*. Available at http://www.nytimes.com/2007/03/28/arts/design/28cnd-pritzker.html?pagewanted=2/.

Putnam, Hilary. 1981. *Reason, Truth, and History*. Cambridge: Cambridge University Press.

Quenqua, D. 2012 (December 4). "Pushing Science's Limits in Sign Language Lexicon." *New York Times*, D1.

Quenqua, D. 2014 (October 11). "Is E-Reading to Your Toddler Story Time, or Simply Screen Time?" *New York Times*, A1.

Quenqua, D. 2014 (October 17). "Quality of Words, Not Quantity, Is Crucial to Language Skills, Study Finds." *New York Times*, A22.

Rasmussen, Majken K., Esben W. Pedersen, Marianne G. Petersen, and Kasper Hornbæk. 2012. "Shape-Changing Interfaces: A Review of the Design Space and Open Research Questions." In *Proceedings of the CHI' 12, the ACM Conference on Human Factors in Computing Systems*, 735–744. Austin, Tex.: ACM.

Read, Gray. 2005. "Theater of Public Space: Architectural Experimentation in the Théâtre De L'espace (Theater of Space), Paris 1937." *Journal of Architectural Education* 58 (4): 53–62.

Read, J., and P. Markopoulos. 2011. "C15: Evaluating Children's Interactive Technology." Course notes distributed at CHI '11, the ACM Conference on Human Factors in Computing Systems, Vancouver, British Columbia, May 7–12, 2011.

Read, R., and T. Belpaeme. 2010. "Interpreting Non-Linguistic Utterances by Robots: Studying the Influence of Physical Appearance." In *Proceedings of AFFINE '10, the 3rd International Workshop on Affective Interaction in Natural Environments*, 65–70. New York: ACM.

Rinaldi, C. 2005. *In Dialogue with Reggio Emilia*. London: Routledge.

Rittel, H. W. J., and M. M. Webber. 1973. "Dilemmas in a General Theory of Planning." *Policy Sciences* 4 (2): 155–169.

Rockwell, Anne F., and Megan Halsey. 2004. *Four Seasons Make a Year*. New York: Walker.

Rosenblatt, Louise. 1978. *The Reader, the Text, the Poem: The Transactional Theory of the Literary Work*. Carbondale: Southern Illinois University Press.

Rossi, Aldo. 1981. *A Scientific Autobiography*. Cambridge, Mass.: MIT Press.

Rutkowski, L. 1982. "On Bayes Risk Consistent Pattern Recognition Procedures in a Quasi-Stationary Environment." *IEEE Transactions on Pattern Analysis and Machine Intelligence* 4 (1): 84–87.

Rykwert, Joseph. 1992. "Organic and Mechanical." *RES* (Autumn): 11–18.

Sadler, Simon. 2005. *Archigram: Architecture without Architecture*. Cambridge, Mass.: MIT Press.

Saval, Nikil. 2014. *Cubed: A Secret History of the Workplace*. New York: Doubleday.

Schafer, George J., Keith Evan Green, Ian D. Walker, Susan King Fullerton, and Elise Lewis. 2014. "An Interactive, Cyber-Physical Read-Aloud Environment: Results and Lessons from an Evaluation Activity with Children and Their Teachers." In *Proceedings of DIS 2014, the ACM Conference on the Design of Interactive Systems*, 865–874. New York: ACM.

Searle, John. 1980. "Minds, Brains and Programs." *Behavioral and Brain Sciences* 3 (3): 417–457.

Sedgwick, Andy. "Would the Pompidou Get Built Today?" Available at http://thoughts.arup.com/post/details/290/would-the-pompidou-get-built-today/.

Sellen, Abigail J., and Richard H. R. Harper. 2002. *The Myth of the Paperless Office*. Cambridge, Mass.: MIT Press.

Sendak, Maurice. 1991. *Where the Wild Things Are*. New York: HarperCollins.

Sellaturay, Senthy V., Raj Nair, Ian K. Dickinson, and Seshadri Sriprasad. 2012. "Proteus: Mythology to Modern Times." *Indian Journal of Urology* 28 (4): 388–391.

Shafer, S. 2000. "Ten Dimensions of Ubiquitous Computing." In *Managing Interactions in Smart Environments*, ed. P. Nixon, G. Lacey, and S. Dobson, 5–16. London: Springer-Verlag.

Short, K. G., G. Kauffman, and L. H. Kahn. 2000. "'I Just Need to Draw': Responding to Literature across Multiple Sign Systems." *Reading Teacher* 54 (2): 160–171.

Shweder, Richard A. 1994 (February 20). "Keep Your Mind Open (and Watch It Closely, Because It's Apt to Change," [Book review of *the Protean Self: Human Resilience in an Age of Fragmentation* by Robert Jay Lifton]. *New York Times*, 16.

Sipe, L. R. 2008. *Storytime: Young Children's Literary Understanding in the Classroom*. New York: Teachers College Press.

Snow, C., M. S. Burns, and P. Griffin, eds. 1998. *Preventing Reading Difficulties in Young Children*. ERIC Document Reproduction Service No. ED 416 465. Washington, D.C.: National Academy Press.

Someya, T. 2013. "Building Bionic Skin." *IEEE Spectrum* 50 (9): 50–56.

Spong, Mark W., Seth Hutchinson, and M. Vidyasagar. 2006. *Robot Modeling and Control*. Hoboken, N.J.: John Wiley & Sons.

Stappers, P. J. 2006. "Designing as a Part of Research." In *Design and the Growth of Knowledge: Best Practices and Ingredients for Successful Design Research*, ed. R. van der Lugt and P. J. Stappers, 13–17. Delft: ID-Studiolab Press.

Stephens, M. "Tame the Web: Libraries and Technology." Available at http://tametheweb.com.

Streitz, N. A., P. Tandler, C. Müller-Tomfelde, and S. Konomi. 2001. "Roomware: Toward the Next Generation of Human-Computer Interaction Based on an Integrated Design of Real and Virtual Worlds." In *Human-Computer Interaction in the New Millennium*, ed. J. Carroll, 553–578. Boston: Addison-Wesley.

"Robots Make Better Employees Than People, America's Business Leaders Say." Available at http://www.hngn.com/articles/41898/20140909/americas-business-leaders-hire-robots-over-people-survey-finds.htm.

Swyngedouw, Erik. 2006. "Metabolic Urbanization: The Making of Cyborg Cities." In *In the Nature of Cities: Urban Political Ecology and the Politics of Urban Metabolism*, ed. Nik Heynen, Maria Kaika, and E. Swyngedouw, 21–40. New York: Routledge.

Syrdal, D. S., K. Dautenhahn, K. L. Koay, and M. L. Walters. 2009. "The Negative Attitudes Towards Robots Scale and Reactions to Robot Behaviour in a Live Human-Robot Inter- Action Study." Paper presented at the Proceedings on New Frontiers in Human-Robot Interaction, AISB 2009 Convention, Edinburgh, Scotland, April 6–9, 2009.

Tafuri, Manfredo. 1976. *Architecture and Utopia: Design and Capitalist Development*. Cambridge, Mass.: MIT Press.

Teale, W. H. 2003. "Reading Aloud to Young Children as a Classroom Instructional Activity: Insights from Research and Practice." In *On Reading Books to Children: Parents and Teachers*, ed. A. van Kleeck, S. A. Stahl, and E. B. Bauer, 114–139. Mahwah, N.J.: Erlbaum Associates.

"The Turing Test, 1950." Available at www.turing.org.uk/scrapbook/test.html.

Threatt, A. L., J. Merino, K. E. Green, I. D. Walker, J. O. Brooks, S. Ficht, R. Kriener, 2012. "A Vision of the Patient Room as an Architectural-Robotic Ecosystem." In *IROS 2012, the IEEE/RJS International Conference on Intelligent Robots and Systems*, 3223–3224. New York: IEEE.

Threatt, A. L., K. E. Green, J. O. Brooks, J. Merino, and I. D. Walker. 2013. "Design and Evaluation of a Nonverbal Communication Platform for Human-Robot Interaction." In *Proceedings of HCI International 2013*, ed. N. Streitz and C. Stephanidis, 505–513. Springer Lecture Notes in Computer Science. Heidelberg: Springer.

Threatt, A. L., J. Merino, K. E. Green, I. D. Walker, J. O. Brooks, and S. Healy. 2014. "An Assistive Robotic Table for Older and Post-Stroke Adults: Results from Participatory Design and Evaluation Activities with Clinical Staff." In *Proceedings of CHI '14, the ACM Conference on Human Factors in Computing Systems*, 673–682. Toronto: ACM.

Trevelyan, James. 1999. "Redefining Robotics for the New Millennium." *International Journal of Robotics Research* 18: 1211–1223.

Tschumi, Bernard. 1994. *Architecture and Disjunction*. Cambridge, Mass.: MIT Press.

Tschumi, Bernard. 1994. *Event-Cities: Praxis*. Cambridge, Mass.: MIT Press.

Tufekci, Zeynep, and Brayden King. 2014, (December 8). "We Can't Trust Uber." *New York Times*, A27.

Turkle, Sherry. 2011. *Alone Together: Why We Expect More from Technology and Less from Each Other*. New York: Basic Books.

Umaschi, M., and J. Cassell. 1997. "Soft Toys with Computer Hearts: Building Personal Storytelling Environments." In *Proceedings of CHI 1997, the ACM Conference on Human Factors in Computing Systems, Extended Abstracts*, 20–21. New York: ACM.

UNESCO. "The Global Literacy Challenge." Available at unesdoc.unesco.org/images/0016/001631/163170e.pdf.

Uras, S., D. Ardu, and M. Deriu. 2012. "Do Not Judge an Interactive Book by Its Cover: A Field Research." In *Proceedings of the 2012 Conference on Advances in Mobile Computing and Multimedia*, 17–20. New York: ACM.

"Urban Acupuncture." Available at http://en.wikipedia.org/wiki/Urban_acupuncture/.

U.S. Department of Education. "National Education Technology Plan: Skills, Education & Competitiveness." Available at www.p21.org/storage/documents/21st_century_skills_education_and_competitiveness_guide.pdf.

U.S. National Science Foundation. "Cyber-Human Systems." Available at www.nsf.gov/cise/iis/chs_pgm13.jsp.

Varshney, U. 2003. "Pervasive Healthcare." *Computer* 36 (12): 138–140.

Vattimo, Gianni, Pier Aldo Rovatti, and Peter Carravetta. 2012. *Weak Thought*. SUNY Series in Contemporary Italian Philosophy. Albany, N.Y.: State University of New York Press.

Venturi, Robert. 1966. *Complexity and Contradiction in Architecture*. New York: Museum of Modern Art.

Vidler, Anthony. 1992. *The Architectural Uncanny: Essays in the Modern Unhomely*. Cambridge, Mass.: MIT Press.

Vitruvius, Marcus Pollio. *De Architectura (Ten Books)*. Trans. F. Granger. Cambridge, Mass.: Harvard University Press, 1931.

von Neumann, John. 1963. "The General and Logical Theory of Automata." In *John Von Neumann: Collected Works*, ed. A. H. Taub, 2070–2098. Elmsford, N.Y.: Pergamon.

Vygotsky, L. S. 1998. "Imagination and Creativity in the Adolescent." In *The Collected Works of L.S. Vygotsky*, ed. R. W. Rieber, 151–166. New York: Plenum.

Weiner, Norbert. 1949. "The Machine Age." Excerpt in Markoff, John. 2013 (May 21). "In 1949, He Imagined an Age of Robots." *New York Times*, D8.

Weiser, M. 1991. "The Computer for the 21st-Century." *Scientific American* 265 (3): 94.

Weschler, Lawrence. 2014 (November 26). "Theo Jansen's Lumbering Life-Forms Arrive in America." *New York Times Magazine*. Available at http://www.nytimes.com/2014/11/30/magazine/theo-jansens-lumbering-life-forms-arrive-in-america.html?_r=0/.

Wigdor, D., C. Shen, C. Forlines, and H. Balakrishnan. 2005. "Effects of Display Position and Control Space Orientation on User Preference and Performance." In *Proceedings of CHI '06, the ACM Conference on Human Factors in Computing Systems*, 309–318. New York: ACM.

Wigley, Mark. 1999. "Recycling Recycling." In *Eco-Tec: Architecture of the in-Between*, ed. Amerigo Marras, 38–49. New York: Princeton Architectural Press.

Wilson, Edward O. 1984. *Biophilia*. Cambridge, Mass.: Harvard University Press.

Wingfield, Nick, and Nick Bilton. 2012 (November 1). "Apple Shake-up Could Lead to Design Shift." *New York Times*, B1.

Winograd, Terry. 1996. *Bringing Design to Software*. Reading, Mass.: Addison-Wesley.

Wiseman, A. 2011. "Interactive Read Alouds: Teachers and Students Constructing Knowledge and Literacy Together." *Early Childhood Education Journal* (38): 431–438.

Yanik, P. M., J. Merino, A. L. Threatt, J. Manganelli, J. O. Brooks, K. E. Green, and I. D. Walker. 2014. "A Gesture Learning Interface for Simulated Robot Path Shaping with a Human Teacher." *IEEE Transactions on Human-Machine Systems* 44 (1): 41–54.

Zimmerman, J., J. Forlizzi, and J. Evenson. 2007. "Research through Design as a Method for Interaction Design Research in HCI." In *Proceedings of CHI '07, the ACM Conference on Human Factors in Computing Systems, Extended Abstracts (Work in Progress)*, 493–502. New York: ACM.

Ziola, R. 2006. "My M D E: Configuring Virtual Workspace in Multidisplay Environments." In *Proceedings of CHI '05, the ACM Conference on Human Factors in Computing Systems, Extended Abstracts (Work in Progress)*, 1481–1486. New York: ACM.

INDEX

Note: Page numbers ending in an italic *f* indicate figures.

2001: A Space Odyssey, 135

Action possibilities, 29
Active rehabilitative robots, 118
Actor-network theory (ANT), 14
Adaptable homes, 85–86
Adaptation to flux, 34
Administrative scalability, 191–193
Aesthetic (formal) dimension, resistance to reconfigurable environments, 26, 29
Aging-in-place assistance, 85, 90, 197. See also home+
AHA (Augmented Human Assistance) project, 199–200
Aiello, Leslie C., 38–39
Alberti, Leon Battista, 26, 29
Alexander, Christopher
   compressed patterns, 45, 68
   early influence on reactive physical environments, 10
   A Pattern Language, 10, 13–14
   pattern recognition, 47
   Sitting Circle pattern, 11*f*
All Watched Over by Machines of Loving Grace, 203
Allen, Stan, 182–183
"Alternate Endings: Using Fiction to Explore Design Futures," 21
Ambient intelligence, 47
Ambiènte workflow configurations, 45–46
Animated Work Environment (AWE). See AWE (Animated Work Environment)
ANT (actor-network theory), 14
Anticipating process, 48
Antle, Alissa, 138, 141
Antón, Susan C., 38–39
Appropriating process, 48
Archigram's Plug-in-City, 174–175

Architects
   of architectural robotics, desired virtues of, 18
   cultivating, 134
"The Architectural Relevance of Cybernetics," 9–10, 14
Architectural robotic library, 186
Architectural robotics
   computer component, 173
   cost, 201
   physical computing, 173
   physical scale, 15–17, 19–21
   scalability, 191–193
   as a Third Place, 201–202
   typologies of (*see* Distributed environments; Reconfigurable environments; Transfigurable environments)
   visual interaction, 181–182
Architectural robotics, future of. See also AWE (Animated Work Environment); home+
   administrative scalability, 191–193
   architectural robotic library, 186
   architectural roots, 173–180
   ARM-N (architectural robotics mesh network), 185–193, 187*f*, 188*f*
   Chinese room problem, 183–184
   CHS (cyber-human system), 173
   cognitive rooms, 184
   computer component, 173
   connectivity conditions, 186, 187*f*
   conscious understanding in machines, 183–184
   cost, 201
   CPS (cyber-physical systems), 173, 192–193
   cyberPLAYce, 185
   definition, 173–174

digital fabrication with robots, 173
distributed cognitive environments, 183–185
dynamic partnerships, 184
ecological design, 201
environment component, 173
functional openness, 192
future of architectural robotics, 185
a garden of technology, 201
geographic scalability, 191–193
goals, 180
habit and habitation, 181–182
heterotropes, 190–191
human component, 173
human-robot interaction, 173
ICT (information and communications technology), 180
ingredients of, 173–180
instrumental scalability, 191–193
intelligent buildings, 173
intelligent environments, 173
Internet of Things, 188–189, 200–202
loading scalability, 191–193
physical computing, 173
physical scale, 181, 201
scalability, 191–193
structural openness, 192
tactile appropriation, 182
tactile interaction, 181–182
tangible computing, 173
technology and the humanities, 193–196
as a Third Place, 201–202
Turing test for machines, 183–184
unconscious intimacy, 182
urban acupuncture, 190–191
visual interaction, 181–182
weak architecture, 190
Architectural robotics, research. See also AWE (Animated Work Environment), researching; Workflow for the twenty-first century, researching
design exemplars, 17 (see also AWE [Animated Work Environment]; Distributed environments; home+; Reconfigurable environments; Transfigurable environments)

design precedents, 17
design research, 17–19
desired virtues of designers, 18
grounding, 17
a hybrid of architecture and robotics, 17
ideation, 17–18
iteration, 18
metaphors, role of, 7, 19–21
protean way, 38, 40f
prototypes, 18–19
STEAM (science, technology, engineering, art, and math), 18
storytelling, role of, 21
underconstrained problems, 17–18
wicked problems, 17–18
Architectural robotics mesh network (ARM-N), 185–193, 187f, 188f
Architectural works, early reconfigurable environments
action possibilities, 29
adaptation to flux, 34
Casa Miller, 32f–33f
changeable rooms and furnishings, 32f–33f, 33
coming-into-being of things, 34
historical examples, 29, 30f–31f
a house under construction, 34
Italian Futurism, 34
jack-in-the-box house, 33
Japanese screen houses, 28f, 29
manually reconfigurable house, 33
mechanical affordances, 29
Rietveld Schröder House, 29, 30f–31f
sliding screens, 29, 30f–31f
unfinished designs, 34
Architectural works, resistance to reconfigurable environments
aesthetic (formal) dimension, 26, 29
beautiful architecture, 29
functional dimension, 26, 29
International Style, 29
Italian Rationalists, 29
Modular man, 26, 27f
Renaissance ideals, 29
typological design, 26
Vitruvian man, 26, 27f

Architecture
  both-and vs. either-or, 124–125
  designed by celebrities, 137
  evidence-based design methods, 193–194
  and industrial design, 193–194
  starchitecture, 137, 175
Architecture-as-scaffolding, 174, 176*f*
Aristotle, 75, 77
ARM-N (architectural robotics mesh network), 185–193, 187*f*, 188*f*
Arm-scale gestures, ART (assistive robotic table), 92–100, 93*f*
ART (assistive robotic table). *See also* Rehabilitative robots
  continuum robotic surfaces, 108–113, 112*f*
  description, 87, 89–90
  diagram, 87*f*
  intelligence level, 100–103
  key components, 90–91, 90*f*
  kinematics, 108
ART (assistive robotic table), communicating with
  arm-scale gestures, 92–100, 93*f*
  bi-directional communication, 95–100
  gesture recognition system, 96–100
  mapping sensed gestures to inferred goals, 99
  nonverbal communication, 92–100, 93*f*
  pattern recognition, 97–103
  user communication with ART, 92–100, 93*f*
ART (assistive robotic table), in a clinical environment
  confirmation of clinician's needs, 105–106
  formative evaluation, 106–107
  high-fidelity prototypes, 106–107
  iterative design and prototyping, 106
  needs assessment, 105
  patient personas, 104
  research phases, 105–107
  setting, 103–104, 103*f*
  summative analysis, 107
ART (assistive robotic table), prototyping
  components of, 90*f*
  a fully functional prototype, 89–90, 89*f*
  high-fidelity prototypes, 106–107
  a key component of *home+*, 87*f*
  in a live setting, 106–107
  required characteristics of upper-extremity rehabilitation robots, 118–120
  starting points, 117–120
  therapy surface, 120, 120*f*
ART (assistive robotic table), scenarios for use
  ethnographic investigations, 90, 91*f*
  first meeting with a user, 92
  mutual learning, 92–94
  patient personas, 104
ART (assistive robotic table), therapy surface
  authoritative model, 123–124
  a fuzzy aggregate, 124
  space of flows, 124
  vortical model, 124
ART (assistive robotic table), therapy surface *vs.* AWE display wall. *See also* OctArm
  constant curvature, 112–115, 112*f*
  description, 108–112
  Interactive Wall, 110*f*–111*f*
  invertebrate robots, 110
  kinematic models, 113, 114*f*–117*f*
  one-dimensional bending *vs.* two-dimensional, 112
  a snake-like room, 109*f*
Artificial intelligence, 195
Artificial muscles. *See* McKibben Actuators
Arup, 174
Assisted-care facilities. *See* home+
Assistive technology. *See* home+
Atmosphere, workflow configurations, 45–46
Augmented Human Assistance (AHA) project, 199–200
Autant, Edouard, 8*f*, 9
Authoritative model, ART therapy surface, 123–124
Automation
  definition, 78
  hospital beds, 25

in reconfigurable environments, 25
in the workplace (*see* Workflow for the twenty-first century)
Automatons
   in the architectural process, 78
   definition, 77–78
   first designs of, 78
AWE (Animated Work Environment)
   vs. coMotion, 48, 50
   components, 40*f*
   CSCW (computer-supported collaborative work), 42
   cyber-physical configurations, 47
   description, 23, 24*f*
   design by task analyses, 43
   vs. humanoid robots, 60–61
   vs. industrial robots, 60
   Interactive Workspaces Project, 42
   key physical elements, 42
   manipulatives, 45
   maximum number of users, 43
   motion planning (*see* Kinematics of robotic movement)
   physical configurations, 4*f*, 55–56
   programmable work surfaces, 42–43, 44*f*, 45
   range of workplace activities, 42
   Roomware, 42
   a snake-like room, 109*f*
   standard configurations, 4*f*, 53–58, 54*f*
   tangibles, 45
   trajectory. *See* Kinematics of robotic movement
   a workstation for the twenty-first century, 41
AWE (Animated Work Environment), display wall
   architectural roots, 178–180
   description, 42–43, 44*f*
   kinematics of moving, 64–65, 66*f*, 67
AWE (Animated Work Environment), display wall *vs.* ART therapy surface. *See also* OctArm
   constant curvature, 112–115, 112*f*
   description, 108–112
   Interactive Wall, 110*f*–111*f*
   invertebrate robots, 110
   kinematic models, 113, 114*f*–117*f*
   one-dimensional bending *vs.* two-dimensional, 112
   a snake-like room, 109*f*
AWE (Animated Work Environment), researching
   developing design guidelines, 52–53
   phone surveys, 50–51
   prototyping, 51–52, 52*f*
   task analyses, 50–51
   WoZ (Wizard of Oz) prototyping, 52, 52*f*
AWE (Animated Work Environment), usability testing
   description, 56, 57*f*, 58
   scenario for, 59

Bachelard, Gaston, 21
Beautiful architecture, resistance to reconfigurable environments, 29
Beauty of machines, 196–199
Benjamin, Walter, 180–181
Best practices for hand drawing, 134
Bibliothèque Sainte-Geneviève, 173–174, 178
Bi-directional communication, ART, 95–100
Bilateral rehabilitative robots, 118
Bionic second-skins, 3
Bizarria Fantastica, 79*f*
The book is a room; the room is a book, 141
Books and publications. *See also* Print media
   All Watched Over by Machines of Loving Grace, 203
   "Alternate Endings: Using Fiction to Explore Design Futures," 21
   "The Architectural Relevance of Cybernetics," 9–10, 14
   The Cat in the Hat, 138–139
   The Circle, 124
   "Concrete Rules and Abstract Machines," 125
   Consciousness and the Social Brain, 164, 166
   "The Corruption of Liberalism," 194–195

Cubed, the Secret History of the Workplace, 41
"Designing as a Part of Research," 18
Digital Ground, 102–103
Discourse on the Method, 195
The Dream of the Moving Statue, 77
"East of Planning: Urbanistic Strategies Today," 190
"The Ethical Significance of Environmental Beauty," 135
Four Seasons Make a Year, 159, 161–163
Future of the Future, 80, 198
"General and Logical Theory of Automata," 78
Generation of Animals, 75
Goodnight Moon, 132
Hebdomeros, 74
Henry VI, 37
How We Learn, 143
"The Hunchback of Notre Dame," 137
L'amour fou, 134
"The Machine Age," 21
"Man Plus," 80
Mechanization Takes Command, 81–83
Mille plateau, 123
Notre-Dame de Paris, 137
Operating Manual for the Spaceship Earth, 199
Paradise Lost, 37
A Pattern Language, 10, 11*f*, 13–14, 47, 100
Physics of the Future, 138
The Protean Self: Human Resilience in an Age of Fragmentation, 37, 41
The Psychology of Human-Computer Interaction, 1, 1*f*
A rebours, 132, 134, 196
Robot Futures, 197
The Robot: The Life Story of a Technology, 77
The Singularity is Near: When Humans Transcend Biology, 100
Soft Architecture Machines, 14, 100
"Statues, Furniture and Generals," 74
Survival through Design, 199
The Tale of Peter Rabbit, 138–139
Ten Books of Architecture, 78
"Ten Dimensions of Ubiquitous Computing," 2, 2*f*
Treatise on Architecture, 134
Two Gentlemen of Verona, 37
Underground, 159, 161–163
Where the Wild Things Are, 129, 139
Borradori, Giovanna, 45
Both-and architecture *vs.* either-or, 124–125
Boudoir in the Mansion House, 72*f*
Bracelli, Battista, 79*f*
Brain
  need for variety, 143
  the reading brain, 167
Brand, Stewart, 194
Brautigan, Richard, 203
Breton, Andre, 133, 139
Brooks, Rodney, 100–101, 124–125
Brown, Denise Scott, 124
Building-as-scaffold, 174

Cage, John, 202
Carey, Benedict, 143
Casa Miller, 32*f*–33*f*
Cassell, Justine, 95
Castells, Manuel, 124
The Cat in the Hat, 138–139
Ceci tuera cela (This will kill that), 137, 178–180
Cedric Price's Fun Palace, 174–175
Centre Pompidou, 173–175
Chairs. *See also* Furniture
  conforming to the human body, 81
  in the Middle Ages, 81
  Stelline chair, 198–199
Chang, Gary, 12*f*
Changeable rooms and furnishings, 32*f*–33*f*, 33. *See also* Reconfigurable environments
Chinese room problem, 183–184
CHS (cyber-human system), 173
The Circle, 124
Cocteau, Jean, 71, 73*f*
Cognitive rooms, 184
Coming-into-being of things, 34
coMotion bench, 48, 49*f*, 50

Compressed living space, 13. *See also* Reconfigurable environments
Compressed patterns, 45, 68. *See also* Reconfigurable environments
Computer component of architectural robotics, 173
Computer-controlled homes, conventional vs. reactive environment, 14
Computer-supported collaborative work (CSCW), 42
Computing embedded in living environments. *See* AWE (Animated Work Environment)
Computing on us and inside us, 3
"Concrete Rules and Abstract Machines," 125
Configurations, AWE, 53–58, 54*f*
Connecting process, 48
Connectivity conditions, 186, 187*f*
Conscious understanding in machines, 183–184
Consciousness and the Social Brain, 164, 166
Continuum robotic surfaces, 108–113, 112*f*
"The Corruption of Liberalism," 194–195
Cost of architectural robotics, 201
CPS (cyber-physical systems), 173, 192–193
CSCW (computer-supported collaborative work), 42
Cubed, the Secret History of the Workplace, 41
Cultivating an architect, 134
Cyber-human system (CHS), 173
Cyber-physical configurations in AWE, 47
Cyber-physical environment. *See* AWE (Animated Work Environment)
cyberPLAYce, 185
Cyclograveur, 176*f*–177*f*

da Vinci, Leonardo, 27*f*
Dali, Salvador, 74
de Certeau, Michel, 23, 25
de Chirico, Giorgio, 71, 74, 76*f*
de Meun, Jean, 133–134, 139
de Saint-Phalle, Niki, 175
de Solá-Morales, Ignasi, 190

Degrees of freedom, kinematics of robotic movement, 62, 64
Deleuze, Gilles, 123–125, 186
des Esseintes, Duc Jean, 132–133, 139
Descartes, René, 195–196
Design exemplars for architectural robotics, 17. *See also* Distributed environments; *home+*; Reconfigurable environments; Transfigurable environments
Design guidelines for AWE, 52–53
Design precedents for architectural robotics, 17
Design research. *See* Architectural robotics, research
Designers of architectural robotics, desired virtues of, 18
"Designing as a Part of Research," 18
di Giorgio, Francesco, 27*f*
Differential kinematics of robotic movement, 61
Digital fabrication with robots, 173
Digital Ground, 102–103
Discourse on the Method, 195
Display wall. *See* AWE (Animated Work Environment), display wall
Dissimilarity principle, 45, 47–48
Distributed cognitive environments, 183–185
Distributed environments
  description, 20
  diagram of, 127*f*
  embedded robotics, 126–127
  history of furnishings, 80–83
  mechanized homes, 80–83
  modular robotics, 126
Distributed environments, living rooms. *See also home+*
  in cinema, 71, 73, 73*f*
  description, 126–127
  a historical perspective, 71–80, 72*f*, 73*f*, 79*f*
  homeliness in the fantasy, 71
  La Belle et la Bête, 71, 73*f*
  mechanical contrivances, 78, 79*f*
  metamorphosis, as analogy for building a house, 75, 77

organic contrivances, 78, 79f
A domestic ecosystem, 90
Domestic Transformer pattern, 12f, 13
Dourish, Paul, 138
Dr. Seuss, 138–139
The Draftsman's Contract, 73, 73f
Drawing, best practices for, 134
The Dream of the Moving Statue, 77
Dynamic partnerships between humans and machines, 184

"East of Planning: Urbanistic Strategies Today," 190
Eco, Umberto, 34
Ecological design, 201
Edelman, George, 7
Eggers, Dave, 124
Either-or architecture vs. both-and, 124–125
Embedded robotics, distributed environments, 126–127
Environment component of architectural robotics, 173
Epistemological engines, 7
E-skin, 3
"The Ethical Significance of Environmental Beauty," 135
Evenson, J., 17–18
Evidence-based architecture, 193–194
Exploring process, 48

Feedback control, kinematics of robotic movement, 62
Filarete, 134
Focillon, Henri, 23, 25
Fondation Jérôme Seydoux-Pathé, 175
Fontaine des automates, 173–177, 176f–177f
Forlizzi, J., 17–18
Formal (aesthetic) dimension, resistance to reconfigurable environments, 26, 29
Forms
    as a definition of space, 23
    different states of, 25
    within forms, 23, 25

Forward kinematics of robotic movement, 62
Foucault, Michael, 190
Four Seasons Make a Year, 159, 161–163
Freud, Sigmund, 71
Fuller, Buckminster, 199
Full-exoskeletal rehabilitative robots, 118
Fun Palace, 174–175
Functional dimension, resistance to reconfigurable environments, 26, 29
Functional openness, 192
Furniture
    designed for workflow in the twenty-first century, 42
    history of furnishings, 80–83
    interactive (*see* Interactive furniture)
    mobility, 80–81
    Rocaille style, 81
    Rococo style, 81
    Rudd's table, 45–46, 46f, 81, 82f
    Sirfo table, 198
Furniture, chairs
    conforming to the human body, 81
    in the Middle Ages, 81
    Stelline chair, 198–199
Furniture in the Valley, 76f
Furniture-man, 79f
Future of architectural robotics. *See* Architectural robotics, future of
Future of the Future, 80, 198
A fuzzy aggregate, 124

A garden of technology, 201
Gaudi, Antoni, 10, 11f
Gee, James Paul, 143
"General and Logical Theory of Automata," 78
Generation of Animals, 75
Geographic scalability, 191–193
Gestalt psychology, supporting workflow for the twenty-first century, 47
Gesture recognition systems
    communicating with ART, 96–100
    GNG (growing neural gas) algorithm, 98
    HMMS (hidden Markov models), 98
    mapping sensed gestures to inferred goals, 99

Giedion, Sigfried, 80–83, 82f
GNG (growing neural gas) algorithm, 98
Goals, robotic, 61
Goodnight Moon, 132
Graves, Marianne, 48
Graziano, Michael, 164, 166
Greenway, Peter, 73
Grönvall, Erik, 48
Gross, Kenneth, 77
Grounding, 17
Growing neural gas (GNG) algorithm, 98
Guattari, Félix, 123–125, 186

Habit and habitation, 181–182
Hale, John, 197–198
Hand drawing, best practices for, 134
Harries, Karsten, 135
Hawking, Stephen, 189
Hawkins, Jeff, 101–102, 197
Hayles, N. Katherine, 183–185
HCI (human-computer interaction). *See also* Ubiquitous computing
  vs. architectural robotics, 173
  bionic second-skins, 3
  circa 1983, 1
  circa 2007, 2
  currently, 3–6
  dialogue between environment and inhabitants, 10
  early view of, 1f
  e-skin, 3
  humanoid robotics, 3
  a reactive environment, 10, 11f
  singularity of machines and humankind, 3
  in the Victorian era, 9–10
HCI (human-computer interaction), as theater
  computing embedded in living environments, 3 (*see also* AWE [Animated Work Environment])
  computing on us and inside us, 3
  more of us in computing, 3
  role of metaphors, 7, 19–21
  scene 1, 1
  scene 2, 2
  scene 3, 3–6, 4f, 7

theater as laboratory, 9
theater *vs.* other types of architecture, 7
Health care facilities. *See* home+
Hebdomeros, 74
Henry VI, 37
Heterotropes, 190–191
Hidden Markov models (HMMS), 98
Hirsh-Pasek, Katherine, 164
Hitchcock, Alfred, 182
Hitchcock, Henry-Russell, 29
HMMS (hidden Markov models), 98
Hoffman, E. T. A., 71
home+. *See also* Distributed environments, living rooms
  adaptable homes, 85–86
  aging-in-place assistance, 85, 90
  big vision approach, 86–88
  as distributed cognitive environment, 183–184
  Intelligent Sweet Home, 86
  Paro, the robotic friend, 197
  a robot for living in, 127
  smart homes, 86
  ubiquitous computing technologies, 86
  uses for, 85
home+, components
  early prototyping, 86–88, 87f–88f
  intelligent headboard, 87, 87f
  intelligent storage, 88, 88f
  over-the-bed table (*see* ART [assistive robotic table])
  personal assistant, 88
  sensitive mobile rail, 87–88, 87f
Homeliness in the fantasy, 71
Homes
  adaptable, 85–86
  mechanized, 80–83
Horváth, Imre, 191–192
Hospital beds, as reconfigurable environments, 25
A house under construction, 34
How We Learn, 143
Hugo, Victor, 137, 178–179
Human component of architectural robotics, 173
Human-computer interaction (HCI). *See* HCI (human-computer interaction)

Humanities and technology, 192–196
Humanoid robotics, 3
Humanoid robots *vs.* AWE, 60–61
Human-robot communication. *See* ART (assistive robotic table), communicating with
Human-robot interaction, 173
"The Hunchback of Notre Dame," 137
Hussert, Edmund, 124
Hutchins, Edwin, 183–184
Huysmans, J. K., 132–133

ICT (information and communications technology), 139, 180
I-cubed, 186, 187*f*
Ideation, 17–18
Ihde, Don, 7
Industrial design, and architecture, 193–194
Industrial robots *vs.* AWE, 60. *See also* Workflow for the twenty-first century
Instrumental scalability, 191–193
Intelligent buildings, 173
Intelligent cars, 25
Intelligent environments. *See also* Reactive environment; Robots for living in
  characteristics of, 14–15
  definition, 14
  ingredient of architectural robotics, 173
Intelligent headboard, 87, 87*f*
Intelligent storage, 88, 88*f*
Intelligent Sweet Home, 86
Interactive furniture. *See also* AWE (Animated Work Environment); Furniture
  anticipating process, 48
  appropriating process, 48
  assistive technology (*see* home+)
  coMotion bench, 48, 49*f*, 50
  connecting process, 48
  exploring process, 48
  hospital beds, 25
  interactive bench, 48, 49*f*, 50
  interpreting process, 48
  over-the-bed tables, reconfigurable (*see* ART [assistive robotic table])
  recounting process, 48
  Rudd's table, 45–46, 46*f*, 81, 82*f*
  sense-making processes, 48, 50
Interactive rehabilitative robots, 118
Interactive wallpaper, 129, 131*f*
Interactive Workspaces Project, 42
Interactivity, as reconfigurable environments, 25
International Style, resistance to reconfigurable environments, 29
Internet, and the persistence of print media, 137
Internet of Things
  vs. architectural robotics, 199–202
  definition, 200
  negative effects of, 188–189
Interpreting process, 48
Inverse kinematics of robotic movement, 62
Invertebrate robots, 63*f*
Ishii, H., 141
Island of Sneetches, 147
Italian Futurism, 34
Italian Rationalists, resistance to reconfigurable environments, 29
Iteration, 18

Jabr, Ferris, 166–167
Jack-in-the-box house, 33
Jansen, Theo, 78–80
Japanese screen houses, 28*f*, 29
Jobs, Steve, 196
Johnson, Phillip, 29
Jung, Carl, 37

KAIST (Korea Advanced Institute of Science), 86
Kaku, Michio, 138
Kaltenbrunner, Robert, 190
Kant, Immanuel, 74–75
Kelly, Kevin, 127
Khunrath, Heinrich, 37
KidsRoom, 147
Kinect, 97–98
Kinematics of robotic movement
  ART (assistive robotic table), 108
  definition, 61

degrees of freedom, 62, 64
differential, 61
feedback control, 62
forward, 62
inverse, 62
invertebrate robots, 63f
moving the AWE display wall, 64–65, 66f, 67
path, 62
plan of motion, 62
redundant robots, 64
rigid-linked robots, 62, 63f
robotic goals, 61
trajectory, 62
vertebrate robots, 62, 63f
Kominers, Paul, 191
Korea Advanced Institute of Science (KAIST), 86
Kubrick, Stanley, 135
Kurzweil, Ray, 100

La Belle et la Bête, 71, 73f
Labrouste, Henri, 178
L'amour fou, 134
Langton, Christopher, 197
Lara, Louise, 9
Le Corbusier, 26, 27f, 29
Lefebvre, Henri, 136
Libraries
  architectural robotic, 186
  Bibliothèque Sainte-Geneviève, 173–174, 178
Life-as-it-could-be, 197
Lifton, Robert Jay, 37
Listen Reader, 147
LIT KIT
  evaluating the prototype, 156–158
  evaluation questions, 157–158
  primacy of print, 166–169
  prototyping, 152–153, 154f–155f
  Smileyometer, 158
LIT KIT, testing in the classroom
  applicability across picture-book types, 161–163
  evaluating the prototype, 156–157
  meaning-making, 160–161

open responses from the children, 163–166
picture books, contextualizing language, concepts, and ideas, 159
sample questionnaire, 157–158
Smileyometer, 158
usability, 159–160
vocabulary acquisition, 160–161
LIT ROOM
  augmented books, 147
  the book is a room; the room is a book, 141
  cultivating literacy skills, 143, 145–146, 167–169
  description, 140f, 141, 142f
  interventions in children's library environments, 148
  Island of Sneetches, 147
  KidsRoom, 147
  Listen Reader, 147
  matching words and images, 151f
  mixed-reality books, 147
  Muscle Body, 143, 144f
  pictures of, 140f, 142f, 169f
  read-alouds, 145–148
  reading role models, 145
  SAGE, 147
  sample evaluation questionnaire, 158f
  story rooms, 146–148
  StoryMat, 147–148
  Wonderbook, 147
LIT ROOM, developing
  customizing environmental effects, 152
  prototyping (see LIT KIT)
  scenario: children discovering the LIT ROOM, 148–152
Literacy
  cultivating with LIT ROOM, 143, 145–146, 167–169
  global, 139
  the reading brain, 167–168
  United States, 139
Living rooms. See AWE (Animated Work Environment); Distributed environments, living rooms; home+
Loading scalability, 191–193
Loureiro, Rui, 118–120

"The Machine Age," 21
Machines, beauty of, 196–199
Machines as art, 178
Madeira Interactive Technologies Institute, 199
Mallgrave, Harry Francis, 7, 75, 195–196
"Man Plus," 80
Manipulatives in AWE, 45
Manually reconfigurable house, 33
Mapping sensed gestures to inferred goals, 99
Marinetti, F. T., 34
Matching words and images, 151*f*
Material semiotics, 14
Mau, Bruce, 51
Mazer, Anne, 129–130*f*
McCullough, Malcolm, 90, 102–103, 191
McHale, John, 80, 201
McKibben actuators, 113–117
Mechanical affordances, 29
Mechanical contrivances, 78, 79*f*
Mechanization Takes Command, 81–83
Medieval churches as reconfigurable environments, 25
Mendini, Allesandro, 198–199
Metamechanics, 176, 176*f*–177*f*, 194
Metamorphosis
  as analogy for building a house, 75, 77
  Aristotle on, 75, 77
Metaphors and the evolution of architectural robotics, 7, 19–21
Microsoft Kinect, 97–98
Microsoft "Office," 41
Mille plateau, 123
Miller, Herman, 41
Mitchell, William, 15, 126
Mixed-reality books, 147
Mobility of furniture, 80–81
Modular man, resistance to reconfigurable environments, 26, 27*f*
Modular robotics, distributed environments, 126
Mollino, Carlo, 29, 32*f*–33*f*, 33, 134
Monet, Claude, 165*f*, 166
Moon landing, 135
Moore's law, 100
Moravec, Hans, 60–61, 184

More of us in computing, 3
Motion planning (robotic movement). *See* Kinematics
Multirobotic rehabilitative robots, 118
Mumford, Lewis, 194
Muscle Body, 143, 144*f*
Musk, Elon, 189
Mutualism, 10

Negroponte, Nicholas
  artificial domestic ecosystem, 101
  a domestic ecosystem, 90
  early influence on reactive physical environments, 10
  on gesture recall systems, 96
  on intelligent environments, 14–15
  the LIT ROOM, 139
  measure of intelligence, 100
Newton (personal digital assistant), 98–99
Nocks, Lisa, 77
Nonstationary environments, 48
Nonverbal communication with ART, 92–100, 93*f*
Norman, Donald, 90
Notre-Dame de Paris, 137, 178–179
Nourbakhsh, Illah Reza, 197, 201

OctArm, 112
Office landscape, 41
One Laptop per Child (XO) Project, 14
Operating Manual for the Spaceship Earth, 199
Organic contrivances, 78, 79*f*
Orr, David, 201
Ortega y Gasset, José, 7
Over-the-bed tables, reconfigurable. *See* ART (assistive robotic table)

Palm Computing, 101
The paperless office, 42
Paradise Lost, 37
Parc Güell, 10, 11*f*
Parish-Morris, Julia, 167
Paro (robotic friend), 197
Partial-exoskeletal rehabilitative robots, 118
Pask, Gordon, 9–10

Passive rehabilitative robots, 118
Path, kinematics of AWE movement, 62
Patient personas, evaluating the ART, 104
A Pattern Language, 10, 11*f*, 13–14, 47, 100
Pattern recognition
    ART (assistive robotic table), 97–103
    supporting workflow for the twenty-first century, 45, 47–48
Patterns
    Domestic Transformer, 12*f*, 13
    Sitting Circle, 10, 11*f*
    supporting workflow for the twenty-first century, 47
    Things From Your Life pattern, 13
"People want to be deceived, so deceive them" (*Vulgus vult decipi, ergo decipiatur*), 171
Personal assistant, 88
Pew Research Center, 200–201
Phone surveys, researching AWE, 50–51
Physical computing, 173
Physical configurations of AWE, 55–56
Physical environments defined by mood, 45–46
Physical reconfiguration. *See* Transfigurable environments
Physical scale of architectural robotics, 15–17, 19–21, 181, 201
Physics of the Future, 138
Piano, Renzo, 174–175
Place, a piling up of heterogeneous places, 25
Plan of motion, kinematics of AWE movement, 62
Plug-in-City, 174–175
Poe, Edgar Allen, 71
Potter, Beatrix, 138–139
Potts, Richard, 38–39
Pragmatic liberalism, 194
Print media, persistence of. *See also* Books and publications
    books of childhood, 138–139
    global literacy, 139
    hybrid analog-and-digital systems, 137–138

ICT (information and communications technology), 139 (*see also* LIT ROOM)
    role of the Internet, 137
    succession of mediums, 138
    United States literacy, 139
Programmable work surfaces, AWE, 42–43, 44*f*, 45
The Protean Self: Human Resilience in an Age of Fragmentation, 37, 41
Protean way
    in architectural robotics, 38, 40*f*
    definition, 37–39
    in the human psyche, 37–39
    role in evolution, 38
Proteus
    Greek myth, 35, 36*f*
    in *Henry VI*, 37
    in modern psychology, 37–39
    in *Paradise Lost*, 37
    in *Two Gentlemen of Verona*, 37
    warships named for, 37
Prototyping
    in architectural robotics research, 18–19
    ART (assistive robotic table) (*see* ART [assistive robotic table], prototyping)
    AWE, 51–52, 52*f*
    home+, 87–88, 87*f*–88*f*
    LIT KIT, 152–153, 154*f*–155*f*, 156–158
    WoZ (Wizard of Oz) prototyping, 52, 52*f*
The Psychology of Human-Computer Interaction, 1, 1*f*
Pygmalion, 133–134

Reactive environment. *See also* Intelligent environment; Robots for living in
    vs. conventional computer-controlled homes, 14
    definition, 10
    examples, 11*f* (*see also* AWE [Animated Work Environment])
    thermostat analogy, 14
Read-alouds, LIT ROOM, 145–148
Reading brain, 167
Reading role models, 145

Reality, the illusion of, 171
A rebours, 132, 134, 196
Recommender systems, 195
Reconfigurable environments. See also
 Automation; AWE (Animated Work
 Environment); home+;
 Protean way
 description, 19, 68, 69f
 the resistance of architecture, 26, 29
 in the workplace (see Workflow for the
 twenty-first century)
Reconfigurable environments, early history of
 action possibilities, 29
 adaptation to flux, 34
 Casa Miller, 32f–33f
 changeable rooms and furnishings,
 32f–33f, 33
 coming-into-being of things, 34
 historical examples, 29, 30f–31f
 a house under construction, 34
 Italian Futurism, 34
 jack-in-the-box house, 33
 Japanese screen houses, 28f, 29
 manually reconfigurable house, 33
 mechanical affordances, 29
 Rietveld Schröder House, 29, 30f–31f
 sliding screens, 29, 30f–31f
 unfinished designs, 34
Reconfigurable environments, early
 resistance to
 aesthetic (formal) dimension, 26, 29
 beautiful architecture, 29
 functional dimension, 26, 29
 International Style, 29
 Italian Rationalists, 29
 Modular man, 26, 27f
 Renaissance ideals, 29
 typological design, 26
 Vitruvian man, 26, 27f
Reconfigurable environments, examples.
 See also AWE (Animated Work
 Environment)
 hospital beds, 25
 intelligent cars, 25
 interactivity, 25
 medieval churches, 25

Rudd's table, 45–46, 46f
Recounting process, 48
Redundant robots, 64
Reggio Emilia Approach
 description, 138
 influence on the LIT ROOM, 141
Rehabilitative robots. See also ART
 (assistive robotic table)
 active, 118
 bilateral technology, 118
 design characteristics, 118–120
 evaluating, 120, 120f
 full-exoskeletal technology, 118
 interactive, 118
 multirobotic technology, 118
 partial-exoskeletal technology, 118
 passive, 118
 unilateral technology, 118
 wire-based technology, 118
Renaissance ideals, resistance to reconfigurable environments, 29
Research
 architectural robotics. See Architectural
 robotics, research; AWE (Animated
 Work Environment), researching
 workflow for the twenty-first century.
 See Workflow for the twenty-first
 century, researching
Responsive environment, 14. See also
 Reactive environment
Richland Library, 145, 163–166. See also
 LIT KIT; LIT ROOM
Rietveld Schröder House, 29, 30f–31f
Rigid-linked robots, 62, 63f
Rittel, H. W. J, 17
A robot for living in, 127
Robot Futures, 197
The Robot: The Life Story of a Technology, 77
Robotic display wall. See AWE
 (Animated Work Environment),
 display wall
Robotics
 embedded robotics, distributed
 environments, 126–127
 (see also AWE [Animated Work
 Environment]; home+)
 goals of, 61

modular robotics, distributed environments, 126
Robots
    definition, 62, 64
    humanoid, 79f
    measuring intelligence level, 100–103
    redundant, 64
    rigid linked, 62, 63f
    soft robotics, 121–123, 121f–122f
    vertebrate, 62, 63f
Robots for living in, definition, 15–16
Robots in the workforce. *See also* Workflow for the twenty-first century
    digital fabrication with robots, 173
    expanding mobility, 61
    humanoid, 60
    increasing compliance, 61
    industrial, 60
    kinematics, 61–62, 63f, 64–65, 66f, 67
    Moravec's paradox, 61
    role of structure in the environment, 61
    supporting twenty-first-century workflow, 59–61
    survey of Harvard Business School alumni, 58
    training the robots, 61
Rocaille style furniture, 81
Rococo style furniture, 81
Roger C. Peace Rehabilitation Hospital, 103
Rogers, John, 3
Rogers, Richard, 174–175
Roomware, 42
Rossi, Aldo, 7, 9
Rudd's table, 45–46, 46f, 81, 82f
Rutkowski, Leszek, 47–48

SAGE, 147
The Salamander Room, 129–130f
Saval, Nikil, 41
Scalability of architectural robotics
    administrative, 191–193
    future of architectural robotics, 191–193
    geographic, 191–193
    instrumental, 191–193
    loading, 191–193

Scale of architectural robotics, 15–17, 19–21, 181, 201
Scenarios for usability testing
    AWE, 59
    LIT ROOM, 148–152
Scenarios for usability testing, ART
    ethnographic investigations, 90, 91f
    first meeting with a user, 92
    mutual learning, 92–94
    patient personas, 104
Science, technology, engineering, art, and math (STEAM), 18
Searle, John, 183–184
Sedgwick, Andy, 174–175
Sendak, Maurice, 138–139
Sense-making processes, 48, 50
Sensitive mobile rail, 87–88, 87f
Shafer, Steve, 2
Shweder, Richard, 41
Singularity
    of machines and humankind, 3, 100–101
    of natural environment and intelligent environment, 101–103
The Singularity is Near: When Humans Transcend Biology, 100
Sirfo table, 198
Sitting Circle pattern, 10, 11f
Size of architectural robotics. *See* Physical scale
Sliding screens, 29, 30f–31f
Smart homes, 86
Smileyometer, 158
Snake-like room, 109f
Soft Architecture Machine Group, 139
Soft Architecture Machines, 14, 100
Soft robotics, 121–123, 121f–122f
Someya, Takao, 3
Space
    defined by form, 23
    of flows, 124
Stappers, Peter Jan, 18
Starchitecture, 137, 175
"Statues, Furniture and Generals," 74
STEAM (science, technology, engineering, art, and math), 18
Stelline chair, 198–199

Story rooms, 146–148
StoryMat, 147–148
Storytelling, in architectural robotics research, 21
Stravinsky, Igor, 175
Stravinsky Fountain, 175–177
Structural openness, 192
"Study of a Figure Outdoors," 165*f*, 166
Survival through Design, 199

Tactile appropriation, 182
Tactile interaction with architectural robotics, 181–182
The Tale of Peter Rabbit, 138–139
Tangible computing, 173
Tangibles, AWE, 45
Task analyses
  AWE, 50–51
  designing AWE, 43
Technology and the humanities, 192–196
Ten Books of Architecture, 78
"Ten Dimensions of Ubiquitous Computing," 2, 2*f*
Theater, role in evolution of HCI. *See* HCI (human-computer interaction), as theater
Théâtre de l'espace, 8*f*, 9
Thermostat analogy for reactive environment, 14
Third Place, 201–202
This will kill that (*Ceci tuera cela*), 137, 178–180
Tinguely, John, 175–178, 176*f*–177*f*, 194
Toy Story, 73
Trajectory (of robotic movement). *See* Kinematics
Transfigurable environments. *See also* AWE (Animated Work Environment); home+; LIT ROOM
  best practices for hand drawing, 134
  creating a space, 135–136
  cultivating an architect, 134
  description, 20, 170–171
  diagram of, 170
  first Moon landing, 135
  the illusion of reality, 171
  interactive wallpaper, 129, 131*f*
  physical reconfiguration, 129–135, 130*f*, 131*f*
  a portal to elsewhere, 129–135
  The Salamander Room, 129–130*f*
Treatise on Architecture, 134
Turing test for machines, 183–184
Turkle, Sherry, 197
Two Gentlemen of Verona, 37
2001: A Space Odyssey, 135
Typological design, resistance to reconfigurable environments, 26
Typologies of architectural robotics. *See* Distributed environments; Reconfigurable environments; Transfigurable environments

Ubiquitous computing, 2, 86. *See also* HCI (human-computer interaction)
Unconscious intimacy, 182
Underconstrained problems, 17–18
Underground, 159, 161–163
Unfinished designs, 34
Unilateral rehabilitative robots, 118
Urban acupuncture, 190–191
Usability testing
  LIT KIT, 159–160
  scenarios for (*see* Scenarios for usability testing)
Usability testing, ART
  ethnographic investigations, 90, 91*f*
  first meeting with a user, 92
  mutual learning, 92–94
  patient personas, 104
Usability testing, AWE
  description, 56, 57*f*, 58
  scenario for, 59
User communication with ART, 92–100, 93*f*
User-programmable (reconfigurable) workstations, 42. *See also* AWE (Animated Work Environment)

Venturi, Robert, 124–125
Vertebrate robots, 62, 63*f*
Victorian era, HCI (human-computer interaction), 9–10
Vidler, Anthony, 71

Visual interaction with architectural robotics, 181–182
Vitruvian man, resistance to reconfigurable environments, 26, 27f
Vitruvius, 26, 27f, 78, 189, 193
Vocabulary acquisition, with LIT KIT, 160–161
von Neumann, John, 78
Vortical model, ART therapy surface, 124
Vulgus vult decipi, ergo decipiatur ("People want to be deceived, so deceive them"), 171
Vygotsky, L. S., 143

Walker, Ian, 112, 121
Weak architecture, 190
Weaning of Furniture Nutrition, 74
Webber, M. M., 17
Weiner, Norbert, 9, 21
Weiser, Mark, 2
Where the Wild Things Are, 129, 139
Wicked problems, 17–18
Wigley, Mark, 201
Wire-based rehabilitative robots, 118
Wonderbook, 147
"The Work of Art in the Age of Mechanical Reproduction," 180–181
Workflow for the twenty-first century. *See also* Robots in the workforce
  demands of, 41
  furniture designed for, 42
  key goals, 42
  Microsoft "Office," 41
  mixed-media use, 42
  vs. nineteenth and twentieth centuries, 41
  office landscape, 41
  the paperless office, 42
  reconfigurable workstations (*see* AWE [Animated Work Environment])
  user-programmable (reconfigurable) workstations, 42 (*see also* AWE [Animated Work Environment])
  work in a non-dedicated workspace, 41
Workflow for the twenty-first century, configurations supporting
  ambient intelligence, 47
  ambiènte, 45–46
  atmosphere, 45–46
  Gestalt psychology, 47
  nonstationary environments, 48
  pattern recognition, 45, 47–48
  patterns, 47
  physical environments defined by mood, 45–46
  principle of dissimilarity, 45, 47–48
  Rudd's table, 45–46, 46f
Workflow for the twenty-first century, researching
  developing design guidelines, 52–53
  phone surveys, 50–51
  prototyping, 51–52, 52f
  task analyses, 50–51
  WoZ (Wizard of Oz) prototyping, 52, 52f
Workplaces, reconfigurable. *See* AWE (Animated Work Environment); *home+*
Workstation for the twenty-first century. *See* AWE (Animated Work Environment)
WoZ (Wizard of Oz) prototyping, 52, 52f

XO (One Laptop per Child) Project, 14

Zimmerman, J., 17–18